The Diamond Formula

The Diamond Formula

Diamond synthesis – a gemmological perspective

A.S. Barnard F.G.A.A. DipDT

OXFORD AUCKLAND BOSTON JOHANNESBURG MELBOURNE NEW DELHI

Butterworth-Heinemann
Linacre House, Jordan Hill, Oxford OX2 8DP
225 Wildwood Avenue, Woburn, MA 01801-2041
A division of Reed Educational and Professional Publishing Ltd

℞ A member of the Reed Elsevier plc group

First published 2000

© Reed Educational and Professional Publishing Ltd 2000

All rights reserved. No part of this publication
may be reproduced in any material form (including
photocopying or storing in any medium by electronic
means and whether or not transiently or incidentally
to some other use of this publication) without the
written permission of the copyright holder except
in accordance with the provisions of the Copyright,
Designs and Patents Act 1988 or under the terms of a
licence issued by the Copyright Licensing Agency Ltd,
90 Tottenham Court Road, London, England W1P 0LP.
Applications for the copyright holder's written permission
to reproduce any part of this publication should be addressed
to the publishers

British Library Cataloguing in Publication Data
A catalogue record for this book is available from the British Library

Library of Congress Cataloguing in Publication Data
A catalogue record for this book is available from the Library of Congress

ISBN 0 7506 4244 0

Typeset by Avocet Typeset, Brill, Aylesbury, Bucks
Printed and bound in Great Britain

Contents

Preface vii
Introduction ix

Part 1 The Story So Far
Chapter 1 Early development   3
Chapter 2 Percy Bridgman and high-pressure physics   9
Chapter 3 Principles of diamond synthesis   15
Chapter 4 Putting the pieces together   25
Chapter 5 Success in diamond synthesis   31
Chapter 6 Birth of a billion-dollar industry   47

Part 2 Synthetic Diamond – Friend or Foe?
Chapter 7 Synthetic diamonds   59
Chapter 8 Gemmology of synthetic diamonds   73
Chapter 9 Scientific testing of synthetic diamonds   81
Chapter 10 Identification of synthetic diamonds   91
Chapter 11 Gemmological testing   97
Chapter 12 Specialized separation instruments   105
Chapter 13 Treated synthetic diamonds   115

Part 3 The New Diamond Makers
Chapter 14 CVD and the new breed of diamond makers   127
Chapter 15 Gemmology and diamond film technology   135
Chapter 16 Other methods of diamond synthesis   143

Epilogue   149
Notes   151
Index   159

Preface

While compiling a gemmological paper on diamond synthesis in 1996, I found a multitude of references and fascinating journal articles outlining every aspect of synthetic diamonds. The information was both rich and varied, taking some months to thoroughly analyse. Some compositions were easily located, others were obscure. This clearly involved a costly exercise for the student.

It was this varied availability that initially persuaded me to consider taking on the compilation of existing research in this field. This inclination was further strengthened by the sudden interest shown in the area of diamond synthesis by associates that were neither gemmologically nor scientifically inclined, when told of my project.

There appears latent interest in this field by members of the wider community, leading to speculation on interest levels within the gemmological and diamond industries. It is concluded that a new reference on the subject may be in order. The concept of making diamonds is one that captures the imagination of almost every one to some degree.

A concise, readable text bringing together all aspects of this exciting subject clearly provides for an obvious need. The goal should encompass the apparent vacuum in readily available gemmological literature on the subject. A prime aim therefore is to provide for the average gemmologist, gemmological or mineralogical student, or other interested people. It will be seen that the gemmologist has a rightful place within this field.

This subject is of great interest to anyone hoping to understand the origin of diamonds and the implication of synthesis for the market. Much of the work in this area has been dependent upon the study of high-pressure physics, coupled with innovations giving rise to the new breakthroughs in diamond synthesis. The field is complex, but the aim is to deal with high-pressure physics to the extent required to direct attention to the subject at hand. Equations have been omitted (with one exception) and technical jargon kept to a minimum. All information relating to the engineering of apparatus and physical chemistry involved in synthesis procedures has also been interpreted in a straightforward manner.

There are many small projects and companies experimenting and producing synthetic diamonds in relatively small quantities. In order to construct a concise report, the majority of references will relate to the largest commercial producers of synthetic diamond and the developmental innovators unless otherwise stated.

I would also like to remind readers that each of the elements and various topics have at some time been the subject of detailed text in their own right, and have been summarized here in order to ensure inclusion. More information can be obtained through the bibliographical sources listed in the Notes section. Where possible, I have tried to utilize references readily available to the average gemmologist or student.

I would like to thank the following people, without whom this book would not have reached its true potential. Marie Milmore and Pam Boiros, for offering professional representation and guidance. Peter Read for his expert scrutiny, editing and allowing me to 'borrow' from his experience. Michael O'Donoghue for his comments and review of the manuscript. Alice Keller of the Gemological Institute of America, for her patience and assistance in arranging for the inclusion of the GIA, *Gems and Gemology* illustrations. Martin Haske of Adamas Gemological Laboratory (AGL) for supplying the spectrophotometer transmittance images for the De Beers synthetic diamonds. David R. Hall and Laurie K. Lilly of the H.T. Hall Foundation for their efforts in preparing and supplying original photographs from the collection of Dr H. Tracy Hall. Martin Cooper and Chris Welbourn of De Beers DTC Physics Department for their suggestions and images. Jun Yamasaki of Sumitomo Electric Industries Ltd, for arranging the supply of the Sumitomo synthetic diamond images. Associate Professor Bernard Pailthorpe, previously of the University of Sydney Plasma Physics laboratory, for allowing me the opportunity of a practical perspective at a crucial stage in my research. Elizabeth Hills of the Gemmological Association of Australia for her assistance in obtaining research materials. Ron and Win Kershaw for their detailed editing and suggestions. Fred Ward for his advice during publication; and finally, Professor Jordan Louviere for inspiration, and Craig Shaw and John Moore for their support.

Introduction

The synthesis of diamonds, the relevant apparatus, and the economic importance placed on the resulting yield, are unique in the broad field of gem synthesis. The race to be the first to produce diamonds in the laboratory has been one of the most highly publicized scientific developments of the twentieth century. Millions of dollars have been invested in research and study of natural diamonds with the aim of producing a synthetic counterpart in an unprecedented manner.

Additionally, scientists were fascinated by the prospect of making diamonds. These modern-day alchemists included Nobel laureates, eccentric inventors, indeed almost every scientist lucky enough to be in possession of a high-pressure device. The research, development and breakthroughs in the field of diamond synthesis are relatively young when compared to the successful synthesis of other gem materials, such as corundum. The first patented process was evolved in the 1950s.

First, we will delve into the history of diamond synthesis, and the early theories, before discussing the principles on which the modern processes rely. The following chapters will cover the advent of success, commercial applications, synthetic diamond products, producers, gemmological properties and identification of synthetic diamonds. Later chapters will briefly discuss diamond synthesis by CVD and application of this technology.

As with the synthesis of numerous 'gemstones', commercial industry has initiated and funded the research with the intention of applying the resulting technology to specialized devices to solve industrial problems. Synthetic diamonds were intended, not for the gem diamonds market, but were born from the need for a reliable supply of super-hard abrasives, machine parts and thermal conductors.

Gemmological interest was largely a by-product initially of relatively little financial consequence. Naturally the success achieved in the 1950s (described in Chapter 5) caused some alarm in the jewellery and gemstone import/export industries. As a result diamond prices during the 1950s radically reflected the public confusion over the effect man-made diamonds could have on wholesale stock pile values, retail stocks in jewellery stores, even the simple engagement ring. As we are now aware, the state of alarm stabilized, but this does not mean that these issues will not resurface in the future.

The major player in the world diamond market, De Beers, was among the companies racing to perfect a method for making diamonds. The feverish international race was manifest in relentless court room battles for control over what is now a multi-billion dollar industry.

An important aspect of this book for the gemmological world will be a discussion of the properties of synthetic diamonds and the analysis of their use in identification of the product. These techniques include advanced analysis and the use of expensive and specialized instrumentation, through to tests that can be easily performed using standard methods and instruments utilized by most gemmologists.

Part 1
The Story So Far

Chapter 1

Early development

The first, and arguably the most important, breakthrough in the successful synthesis of diamonds was the discovery, born from the 1772 theories of French chemist Antoine Lavoisier together with additional experimental proof in 1797 by the English chemist Smithson Tennant, that diamond was in fact a form of carbon. As early as the seventeenth century, however, Robert Boyle had shown that diamond was physically affected by the application of a flame. In 1694 G. Averani and C.A. Targiono (of the Florence Academy in Italy) demonstrated that diamonds actually burn if subjected to appropriate thermal energy.

Later Lavoisier determined that diamonds were formed from carbon by showing that the gas produced when a diamond was burnt was similar in nature to the gas produced when charcoal was burnt. His experiment was performed by sealing a diamond inside an oxygen-filled jar. With the aid of a powerful lens, he focused the rays of the sun into the jar and induced the diamond to burn.

Tennant confirmed this by fusing the diamonds with nitre (potassium nitrate) in oxygen, and identifying the resulting gas as carbon dioxide. In addition he was able to prove that when equal weights of diamonds and charcoal were burnt, the result was equal weights of carbon dioxide.[1]

Until this time many theories had been proposed as to the origin and nature of diamonds, ranging from the notion that they were in fact a form of ice, frozen for a very long time, to the belief that they were pieces of stars fallen to earth.[2]

Prior to 1779, molybdenum sulphide and graphite were confused and consequently thought to be the same substance. When this was disproved by K.W. Scheele it was assumed that graphite must be a carbide of iron. It was not until 1855 when Brodie prepared pure graphite in a specialized furnace in his laboratory that it was recognized that graphite was a genuine allotropic form of carbon.[3]

Many questioned whether diamond could possibly be a form of carbon, when the (third) solid allotrope of carbon, amorphous carbon, had already been established (ironically the same conjecture plagued buckminsterfullerene following its discovery, as to whether it would be recognized as a formal allotrope[4]). It must be pointed out however, that amorphous carbon is not formally considered a mineral.[1] This confusion in the nineteenth century can be understood. All of the physical and visual characteristics of the two substances were at complete odds with each other.

Conjecture continued for some time and was compounded still further when graphite was isolated. Graphite and amorphous carbon are similar, so it seemed reasonable that graphite was the only crystalline allotrope of carbon. How could graphite and diamond be chemically the same? Diamond was the hardest known mineral and graphite one of the most fragile; diamond was transparent and graphite opaque.

It was later proposed that if diamond was indeed carbon, which had a specific gravity of 3.52, while graphite (also known to be carbon) had a specific gravity of 2.24, then obvious differences lay in their relative densities. It was known at the time that substances formed under pressure result in denser media, and from this theory first appeared the concept that diamond must be a highly pressurized form of graphite.[2]

During the nineteenth century most of the studies into the possibility of making diamonds in the laboratory focused on understanding natural diamonds. Scientists looked for clues as to how nature was able to manipulate carbon into the closely packed atomic arrangement of diamond. Natural diamonds were subjected to many laboratory experiments. In the field, geologists and geophysicists delved into the mysteries of the earth. Many distinguished scientists made and broke reputations trying to make diamonds in the late nineteenth and early part of the twentieth century. Here we will briefly look at the work of perhaps the most famous of these early attempts.

It was, however the fundamental idea that diamond was a highly pressurized form of carbon that formed the basis for activities for the first of the serious would-be diamond makers of the late 1800s. In 1878 a Scottish chemist, James Ballantine Hannay, was among the first to attempt to produce diamonds in his albeit crude (by today's standards) laboratory.[2]

Hannay began his career at the Shawfield Chemical Works in Glasgow and later became an assistant lecturer at Owens College in Manchester. He opened his own laboratory with the intention of making diamonds. His predecessors had already isolated the crucial primary ingredients required (carbon and intense pressure), or so he thought. Using this cue, Hannay attempted to put the pieces together. In his considered opinion diamonds could be made when a gas containing carbon and hydrogen (a hydrocarbon) is heated under pressure in the presence of lithium, potassium, sodium or magnesium. The hydrogen would bond with the metals, and the remaining carbon would crystallize, possibly as diamond.[2]

Hannay's theory on the synthesis of diamonds seems inadequate to us, perhaps idealistic. His apparatus consisted of little more than a large ceramic style furnace, and an iron cylinder.[2] He tested his theory with solid phase constituents such as light paraffin, bone oil and metallic lithium.[5] The experiment was performed by placing the volatile carbonaceous compounds in a sealed iron cylinder 10 cm in diameter with a 1.25 cm bore. It was then welded closed using the usual techniques employed at the time.

The cylinder was heated in a furnace for some hours to a high temperature sufficient to cause an explosion estimated to be approximately 7000 atmospheres of pressure by today's standards.[6] Hannay believed that the process of heating the carbonaceous material to a high temperature would cause it to expand within the cylinder, resulting in an internal build up of pressure. The pressure, Hannay thought, would cause the carbon within the compound to be converted to diamond. He would then simply retrieve the synthetic diamonds from the iron cylinder, when cooled, by immersing it in acid. The cylinder would be lost, but the cost would be insignificant when compared with the possible production of diamonds.

Of the approximately 80 experiments[1] conducted by Hannay and his assistants only three cylinders survived the explosions. Some of the explosions were so violent that they destroyed the entire furnace; one such explosion injuring (almost seriously) one of Hannay's assistants![2] The surviving cylinders or fragments thereof, were all subjected to acid and analysed for evidence of diamonds, without success.[6]

It was a long, laborious and dangerous process that maintained Hannay's attention for many years, through many failed experiments. Finally diamond fragments were found after one such experiment.[1] As was the standard procedure of the time (and even now), the tiny crystals were sent for independent identification. One can imagine the feeling of elation, and relief as Hannay sent his 'synthetic' diamonds for verification to the British museum, where they were found in fact to be diamond. Hannay published his success in the *Proceedings for the Royal Society* (1880), and retired from this field of research.[2] Only nine of these diamond fragments have survived to this day.

The professor in charge of the verification activities in the mineralogical department of the British Museum was Nevil Story-Maskelyne. Story-Maskelyne had limited tests at his disposal, and in the end, three tests were performed to verify the samples as diamond. These were a scratch test, in which it was determined that Hannay's diamonds easily scratched sapphire, a mineralogical examination that revealed the cleavage angle of 70.5° (as is consistent with diamond), and an experiment based on the principles of Tennant and Lavoisier – that diamonds burn in oxygen. Hannay's samples passed all of these tests. Hence Story-Maskelyne positively identified the fragments as diamond. He published the results in *The Times* (London) on 20 February 1880.[1]

Unfortunately later analysis with modern microscopy performed by Lonsdale showed the fragments to be natural diamonds and not synthetic, due to the presence of platelet defects not possible in synthetic diamonds even to this day. Subsequent supporting analysis by Collins revealed the inclusion of nitrogen in aggregated concentrations. This was further proved with the use of X-ray diffraction, cathodoluminescence and chemical analysis[6] (more on these techniques in later chapters). Many possible reasons have been suggested, with the most plausible, and widely quoted, being the accidental contamination of the original mixture with natural diamond fragments before experimentation.[6]

The majority of scientists conducting experimental research in this field were known to possess a small collection of natural diamond, purely in order to study it and its physical, optical and mechanical properties. It is feasible that such a sample may have found its way into the mixing pot.

Hannay was not the only eminent scientist to try his hand at diamond synthesis and prematurely claim success. 1906 Noble Prize recipient Dr Ferdinand Frédéric-Henri Moissan began research into diamond synthesis in 1889 using an entirely different approach to Hannay.[2] Moissan's course of action was said to have been inspired by the news of the discovery of diamonds at the Canon Diablo meteorite site.[5] His first step was to design a completely new type of apparatus for the purpose of diamond making (although it was also suitable for studying other materials).

His apparatus was much more complex, and rather less dangerous than that of his predecessor. It consisted of a pair of carbon rods, protruding into a small cavity between two blocks of lime. The external ends of the rods were attached to charged electrodes, to enable electrical heating.[2] The cavity between the blocks was filled with the sample. In Moissan's case the sample was a compound of iron and sugar charcoal.[6]

In theory, an electrical current passed via the rods would heat the sample to unprecedented temperatures. Moissan thought this would be sufficient to enable transition from carbon to a volatile carbon phase without crystalline structure, just waiting to be converted to diamond. He believed that the sample would be effectively heated due to the heat flow from iron to carbon, in a type of thermal gradient. The experiment worked like a charm in one respect. The current was passed at approximately

700 amperes at 120 volts, and, just as anticipated, the sample within the lime crucible was heated to around 3000°C, a temperature that at the time revolutionized the field of temperature chemistry.[1]

The compound resulting from such an experiment would usually cool to form cast iron. Moissan theorized that if plunged in cold water it would cool rapidly from the outside in, resulting in an extreme build up of pressure within the rapidly forming cylinder. He believed this massive build up of pressure inside the cooling iron mass would allow the extra carbon to crystallize at pressures conducive to diamond formation.[6] Later he modified this step of the procedure to increase the rate of cooling. To achieve this he immersed the iron into baths of molten lead.[2] Moissan, like Hannay, conducted dozens of experiments, without much success. Crystals were synthesized, but found not to be diamond.

However upon dissolving the iron in acid after one such experiment, Moissan did liberate tiny octahedral crystals. The crystals extracted were verified externally as being diamonds, as they adopted the octahedral form, scratched other hard minerals and left residual carbon dioxide when burned in oxygen.[2] For many years it was believed that man had successfully duplicated nature's greatest feat. However, following Moissan's death in 1907 it was revealed that the specimens were in fact natural diamonds and not synthetic as believed. The crystals had apparently been planted in the mixture by one of Moissan's assistants who felt pity for the great scientist's repeated failure.[6]

The question of the identity of the crystals that *had* been synthesized during such experiments remained. These experiments have been performed in modern laboratories, using the original set-up used by Moissan, in an attempt to isolate the crystals and analyse their chemical nature. They have been successfully reproduced and found to be silicon carbide. Silicon carbide (like a number of other materials) was synthesized before it was discovered in nature. Following this discovery it was given the name Moissanite, in honour of the great scientist that confused it with diamond so many years ago.[2]

Moissan's theory was published in 1889 and was subsequently adopted by many scientists. Following Moissan's death in 1907 many other scientists claimed to have succeeded. This group of specialists included Sir William Crookes who claimed in 1909, and for many years following, to have succeeded in synthesizing diamonds using the Moissan technique. He later withdrew his claims when it was revealed that the Moissan crystals were not synthetic diamond after all. Crookes also tried his own method, involving the explosion of cordite and carbon in a closed steel vessel. He also claimed success with this method! Others to attempt to replicate the work of Moissan include O. Ruff (1917) and J.W. Hershey (1929).[6]

Both Hannay's and Moissan's methods were repeated numerous times by English engineer Sir Charles Algernon Parsons in an attempt not only to verify the claims of the two scientists before him, but also to perfect the methods for commercial use. Parsons, born in London, was a privileged gentleman who claimed among his accomplishments, the invention of the steam turbine[2] for which he received his knighthood.

With his education, resources and previous successes, Parsons appeared to be the logical candidate to take up the quest for synthetic diamonds, and dedicated thirty years of his life to this cause.[6] He was known for his painstakingly accurate approach and methodical record keeping. All his resulting samples were carefully stored for later analysis by an independent party.[2] In addition to the attempted duplication techniques he tried techniques of his own, including firing bullets into carbon-containing materials, and heating carbon rods immersed in quartz and calcium oxide under pressure.[6]

Unfortunately Parsons was unable to produce any synthetic diamonds by repeating Hannay's method with the utilization of super-heated iron cylinders and carbon. For many years he believed he had succeeded in his quest utilizing Moissan's techniques, as well as a number of his own methods. However as more sophisticated analytical tools were developed, the crystals he had produced proved to be spinel.[2]

One can only speculate as to whether it was the octahedral crystals of spinel that had so effectively fooled Parson's predecessors, in conjunction with the before mentioned silicon carbide, later named Moissanite. The scratch test was employed in early identification; and while spinel (a hardness of 8 on the Mohs scale) does not approach the hardness of diamond (the standard selected to represent 10 on the same scale) it does scratch quartz, an often used comparative material. In 1918 Parsons was reported as saying that he believed that pressures up to 18 000 atmospheres were still not sufficient to produce diamonds. A paper was published in 1928 by C.H. Desch, authorized by Parsons himself, stating that up to that time no diamonds had been made by man in the laboratory.[6] The community as a whole was forced to wonder if the synthesizing of diamonds would ever be achieved.

In hindsight, however, early attempts were on the right track. Their chemistry and applications were quite similar to those used in the synthesis of diamonds today, without approaching the required levels of course. While their principles were correct, that carbon must be subjected to great pressure, the use of heat in both methods was employed to facilitate the pressure required, rather than as a necessary ingredient in its own right. Early scientists failed to realize the importance of simultaneous heat and pressure rather that one begetting the other. The importance of this will be fully discussed in Chapter 4.

Moissan dissolved his carbon in iron, drawn from the association of diamond with iron in nature, not realizing what an important decision he had made toward eventual success (unfortunately not his success!). Failure can best be attributed to the facilities and technology available to these early diamond makers. Their apparatus simply did not allow them to approach the extreme levels of pressure and temperature (or combination of these) required to initiate the transformation of their samples.

If the quest was to continue and improve, then major advancements had to be made involving the technologies that were employed. In addition to this, there was no reliable way of measuring the pressures and temperatures attained at the time, let alone calculate the levels to strive for.

This was in part due to the fact that (at least in the case of Moissan's temperature) new ground was being broken; with every experiment lay the possibility of a new record in high pressure and high temperature physics. This was certainly the case with Percy Bridgman.

Chapter 2

Percy Bridgman and high-pressure physics

Percy Williams Bridgman, the winner of the 1946 Nobel Prize for physics was for most of his career a physicist at Harvard, working in high-pressure physics.[6] He was without doubt a revolutionary leader in this field. Even after retirement, he was persuaded to consult on many projects relating to the synthesis of diamonds.

Born in 1882 in Cambridge, Massachusetts, Bridgman developed an early passion for science. He was a diligent student, although as a youth he was rebellious against the plans and ideals of his father, a journalist and a devout man. His father instilled in him a strong work ethic and a spirit of honesty and of determination. Young Bridgman's devotion to his academic studies was matched by a wide range of extra curricular activities. Though still regarded by his friends as a quiet loner, he was to contribute more to the high-pressure history than anyone before him. He worked on a myriad of projects including the war time Manhattan Project.[2]

Following his formative public school education at Newton, Massachusetts, he undertook tertiary studies at the prestigious Harvard University where he graduated summa cum laude in 1904 with a Bachelor of Arts. He completed his Master of Arts in 1905, was awarded his doctorate in 1908,[6] and continued his career at Harvard progressing from doctorate student to research fellow, and eventually professor emeritus.

While Bridgman appeared happy at Harvard, he had little time and patience for the day-to-day dealings with students, lectures and internal politics. Indeed he seemed to be as unpopular with his graduate students as he was with the undergraduates, although his reputation and first rate course attracted the finest physics applicants. Among his students were the quantum theoretician John C. Slater and J. Robert Oppenheimer, of the Manhattan Project.[2]

During his thesis work (on the effects of pressure on refractive index) in 1905, Bridgman was introduced to the field of pressure physics almost by accident. An explosion in his laboratory was responsible for his stumbling upon a new type of high-pressure apparatus. This small and coincidental occurrence changed the course of his career, and started him on the quest for man-made diamonds. Following the attainment of his doctorate, he began his new course of research into high-pressure and its effects on the phase transition of different materials.[2]

He began work pressurizing a multitude of materials, almost a thousand different types in all. Many experiments resulted in the development of new exotic materials. Whatever substances he could lay his hands on, he pressurized and enthused at the exciting results. He discovered many high-pressure forms of ordinary substances, as yet unknown to science. With each experiment new materials emerged, such as five new high-pressure forms of ice.[2] This accomplishment was not accidental. Years of work and instrumental modification were necessary before this could be achieved.

Initially he had used a crude screw apparatus turned by a six foot wrench,[6] a device he concluded to be unsatisfactory for the tasks that lay before him. He first set about analysing the apparatus employed by the Harvard laboratory, making his own improvements and amendments as he saw necessary. The main components requiring attention were the press (to induce the pressures through external force), the anvil (contacting with the sample) and of course the sample. His aim was to dramatically increase the capacity of the device with regard to the level of pressure attainable, however there were two major problems presenting apparently insurmountable hurdles. First, the press simply could not exert enough pressure on the sample. Second, when it did approach high-pressures, the sample leaked.[2]

This problem of leakage was solved with a simple and yet remarkably effective addition to the device. Bridgman designed an unsupported seal that contained the sample within the apparatus without rupturing during high-pressure application. This seal was composed of a ring of the soft and malleable rock called pipestone[2] which was later replaced by researchers with pyrophyllite.[6] Pyrophyllite (like pipestone), is a light tan coloured rock with a fibrous almost soapy appearance, revealing none of its deformable properties at room pressures.[7]

At high-pressures deformation of this pyrophyllite (and pipestone) is uniform. To some degree, force exerted by the press is counter balanced by a resisting force exerted by the mineral, apparently trying to return to its original state. This produces a self-tightening gasket that does not rupture or break, and effectively prevents leakage.

The invention of the non-extruding gasket opened the door to a new era of high-pressure physics. Bridgman was able to achieve pressures unprecedented at the time. He was the first to isolate a whole new series of high-pressure materials.[2] Unfortunately these new high-pressures were still not enough to convert graphite to diamond, and synthetic diamonds eluded him.

Always striving for higher pressures Bridgman continued to improve his press. In 1935 he developed what is now known as the 'Bridgman anvil', an elegant modification to the mainstream press and anvil design of the time.[2] This design became the preliminary device of the Hall Belt apparatus (outlined in Chapter 5). The Bridgman anvil was based on a simple equation for pressure multiplication.[6] Given that pressure is defined as force per unit area:

$$F = f(A/a) \tag{1.1}$$

where:

f is the force applied to the press
A is the larger area (of the hydraulic press)
a is the smaller area (of the tapered anvil), and
F is the resulting force achieved by the press.

The principle is one adopted by many leaders in the field of high-pressure physics, not only to strive for new levels of high pressure, but also in an attempt to measure the pressures that they have been successful in achieving. The basis rests in the mathematical fact that a force (F) acting on a surface (A) will give a resulting pressure.

However if the same force (F) is focused onto a much smaller area (a), then the resulting pressure will be multiplied. The amount of the increase will be directly proportional to difference between the two surfaces; the size (surface area) of the larger area (*A*) divided by the size (surface area) of the smaller area (*a*). This equation can be verified as being valid using standard dimensional analysis.

A simple example of this is the hole-punch. A fist or fingers alone (at normal distances) can not punch through, say twenty pieces of typing paper for example, even if we punch as hard as possible. Nor can penetration be achieved through a stack of paper lying across two supports. The stack can be deformed, but a clean and neat hole cannot result from these endeavours. However, when a hole-punch is employed, the force of the hand (*f*) on the handle of the device (*A*) is multiplied by the relationship of the larger area of the handle (*A*), to the smaller area of the end of the piston (*a*). The resulting force (*F*) easily passes through the stack of paper. A schematic for this principle applied to the Bridgman Anvil is provided in Figure 2.1.

It is through this device that the relatively lower pressures applied by the hydraulic press alone give rise to the high-pressures required. With the hydraulic press Bridgman was able to apply pressures of 7000 atmospheres. The addition of the non-extruding gasket increased this to around 20 000 atmospheres, and with the tapered anvils this was promoted to around 50 000 atmospheres.[2] These pressures were at the time revolutionary. One of the first tasks following this scientific breakthrough was to develop a way to measure it. We can now discuss with ease the pressures achieved at each milestone on his road to success, however at the time pressures had to be estimated.

Science has developed a number of pressure scales (see Chapter 3), but Bridgman needed a more practical and applicable system to use in his work. The use of metal deformation, mercury weights, and fluid volume changes were widely accepted and applied in the scientific community, however these were not good enough. The perfectionist in him demanded a more accurate and efficient approach – one that could

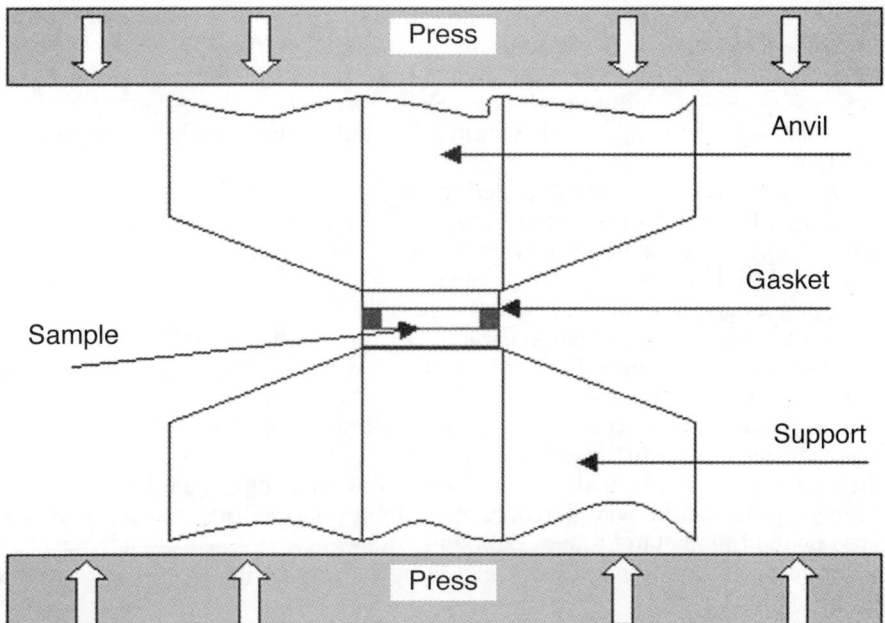

Figure 2.1 The Bridgman anvil, featuring tapered anvil design and non-extruding gasket.

measure the smallest pressure fluctuations, but still capable of measuring the high-pressure levels he was yet to achieve. To this end Bridgman developed his own system based on the decrease in mercury's electrical resistance with pressure. This new mercury resistance gauge method was applauded by the high-pressure community and was used for decades to come.[2]

Bridgman embarked on another series of experiments, as all of the previously tested materials could yield a new phase at the new high-pressures. He made his first headline with the discovery of the five high-pressure solid forms of the most simple and abundant of materials, water. He pressurized the water to 20 000 atmospheres and at different pressures and temperatures documented the results. One form (ice VI) was grasped by the public (wrongly) as a new marvel for commercial industry. Nicknamed 'hot ice' the variety was stable at 95°C, which many people thought meant a solid 'ice' form of hot-water. Sadly hot ice did not hold any new secrets for a hopeful world of industrialists as once the pressure is released the solid immediately melts.[2]

The revolution continued for over 20 years, with Bridgman later being able to achieve pressures of 400 000 atmospheres.[1] A whole new realm of exotic minerals and new phases of existing minerals opened up to Bridgman, however one particular door seemed permanently locked.

Of the many successes achieved by Bridgman, one mineral plagued him with continual failure – graphite. Bridgman was quoted as saying that in graphite he had found the world's greatest spring. No matter how much he squeezed it in his device, it always reverted to graphite when the pressure was removed.[1] While it is not known exactly how many attempts Bridgman made to create diamonds over the course of an illustrious career, he himself admitted that none were successful. It has been said that after each alteration or improvement of the high-pressure device, or the introduction of new componentry, graphite was his first test case.[2]

High-pressure research was a dangerous exercise. Bridgman risked life and limb in his laboratory with exploding cylinders and rupturing apparatus. Time and time again experiments were ruined. It was reported that his laboratory bore the brunt of the high-pressure device's fury, the walls punctured with rather imposing holes made by parts of his apparatus. Sadly on 19 May 1922, research fellow Atherton K. Dunbar and assistant William Connell fell victim to a fatal explosion while pressurizing an oxygen tank.

In 1931 Bridgman published a definitive book called *The Pressure of High-Pressure*, in which a whole chapter devoted to experimental catastrophes listed the 'variety of interesting ways' a device could rupture. These disasters lead him to develop a method of shrink fitting a steel girdle, termed 'massive support' to his central components to reduce the risk of failure.[2]

In January 1941 the Carboloy Division of General Electric together with the Norton and Carborundum companies formed an alliance with the clear objective of succeeding in the quest to make diamonds artificially. It was Percy Bridgman who spearheaded the team. He was given access to the latest super-hard materials with which to craft his devices and all the assistance and resources he required.[2]

After two years with the team, and no successful runs on the board, the project was abandoned. With World War II reaching its final phase, Bridgman's expertise was required on the Manhattan Project. Following the war, in 1946 Bridgman received his Nobel Prize in physics for his work and contributions to the field of high-pressure research.[6]

In 1959 the National Bureau of Standards approached Harvard University with the aim of collating Bridgman's work. The project was funded by the National Science

Foundation in 1961, and undertaken by Bridgman himself. This endeavour encompassed 198 papers, and his last achievement was published in 1964. Upon completing the volumes Bridgman (suffering from inoperable bone cancer) took his own life. Thus ended an important chapter in scientific history.[2] Before he died however Percy Bridgman did make diamonds. He was invited to join the company that eventually was successful, as a sign of respect and no doubt as a way of thanking Bridgman for starting them on the right road to discovery.[2]

The legacy begun by Bridgman and the quest for synthetic diamonds continued. We will now discuss each of the ingredients required for diamond synthesis and how they fit together.

Chapter 3

Principles of diamond synthesis

Three of the key ingredients required for the synthesis of diamonds are:

- carbonaceous material
- high pressure, and
- high temperature.

These will be discussed separately.

Carbon

Carbon, the element of diamonds, has an atomic weight of 12.01115 and an atomic number of 6, and is denoted by the letter C, being in the group 4 (or IVa) of the periodic table. Carbon has an electron configuration of 2-4, meaning it has two electrons in the innermost shell (the K shell) and four in the outer shell (the L shell). Carbon has a valency of 4, allowing it to form covalent bonds with four other carbon atoms. Carbon is known to exist in four autonomous states, the two crystalline solids diamond and graphite, amorphous carbon (including the gas phase),[3] and the family of fullerenes (the most famous of which is the C_{60} buckminsterfullerene) discovered in 1985.[2] Carbon has a high propensity to form molecules, organic or inorganic.

Carbon is one of the most commonly occurring elements on the planet although it makes up only about 0.025 per cent of the earth's crust.[8] It occurs there mostly in the form of carbonates but is also one of the building blocks of living matter, including humans. Carbon constitutes more than 50 per cent of our dry body weight.[9] The total carbon pool, distributed among organic and inorganic forms, is estimated at about 49 000 gigatonnes (1 gigatonne equals 10^9 tonnes). Fossil carbon accounts for 22 per cent of the total pool. The oceans contain 71 per cent of the world's carbon, mostly in the form of bicarbonate and carbonate ions. An additional 3 per cent is in dead organic matter and phytoplankton. Ecosystems, in which forests are the main reservoir, hold about 3 per cent of the total carbon.[8]

As we have discussed briefly in Chapter 1, three of the allotropes of carbon were established by the middle of the nineteenth century; however, the study of carbon originated as far back as the third century AD when carbon was used as a dye. For many centuries few attempts were made to diversify the applications for carbon in either industrial or chemical contexts due to its apparent inability to react with other elements. The term used to describe this property of carbon (and other elements with this property) is 'chemically inert'. As further study revealed the nature of carbon and its chemical behaviour, more applications were developed. In the 1950s carbon in the form of (synthesized) graphite was used in atomic reactors. The invention of carbon fibre technology followed.[4]

Today, carbon technology is a thriving industry. Examples of the diverse applications of carbon are contained throughout this text. Carbon forms the lead in our pencils, is a major constituent in cast iron and many super-hard materials such as silicon carbide and tungsten carbide. Synthetic diamonds and diamond films (to be discussed later in Part 3) are fast becoming important commodities. In addition to the commercial use of carbon, many disciplines of science have arisen over the decades, dedicating themselves completely to the study of carbon.

Typically inorganic carbonaceous materials are employed in diamond synthesis, but this need not be the case. Almost any substance rich in carbon can be converted to diamond if subjected to the optimum conditions for each particular substance. These substances include carbon gases such as methane and propane, inorganic carbon containing materials such as coke, graphite, coal, buckminsterfullerenes (see discussions in Part 3), man-made plastics and polymers, and finally organic substances such as sugar or paraffin wax.[6] Hence the study of inorganic chemistry has a bearing on the synthesis of diamonds.

The study of organic chemistry is in many respects the study of carbon, as live molecules are carbon based.[9] Combustion science is a specialist field of organic chemistry that deals completely in the study of elementary carbon and the combustion of hydrocarbon fuels.[4] Carbon science has been described as a rather colourful mix of many fields considered to be specialties in their own right. The science of carbon combines aspects of physics, chemistry, chemical engineering, geology, astrophysics and fluid dynamics, all bundled together in the search for the understanding of one of the most fundamental elements.

The most recent addition to the realm of carbon science is the study of fullerenes. The famous buckminsterfullerene was discovered by a group of scientists at Rice University in Houston, Texas. The team originally set out to compile data on clustering carbon molecules, and to establish their existence in space. The team was headed by Harry Kroto, Rick Smalley and Robert Curl who worked in conjunction with graduate students Jim Heath, Sean O'Brien, Yuan Lui and Qingling Zhang.[4]

As we can see from Figure 3.1, buckminsterfullerene is a 60-atom carbon molecule (written as C_{60}) in the form of a truncated icosahedron. Each junction in the diagram is the site for a carbon atom.[4]

Some scientists believe that the natural diamonds determined to be of eclogitic origin derive their carbon from the remains of organic carbon containing life forms, subducted to great depths by the process of plate tectonics.[10] This is indicated by the inordinately larger quantities of the carbon-13 isotope often associated with organic materials found in the lattice of these particular natural diamonds. The utilization of organic carbons (of either isotope), in effective non-diamond carbon to diamond carbon transition, has also been achieved in the laboratory. Peanuts yield diamond with 60 per cent conversion efficiency in around five minutes (at 2000°C and 140 000 atmospheres).[6]

Carbon-13 is a heavier isotope of carbon, with an extra neutron in the atomic nucleus. This carbon isotope accounts for approximately 1 per cent of carbon atoms and is written as C^{13}. The remaining 99 per cent of carbon atoms are the 'standard' carbon-12 variety, with equal number of protons and neutrons in the nucleus (or the elusive carbon-14 used for carbon dating, and carbon-11).[4] The significance of this will become apparent in Chapter 7.

As it is graphite that is favoured by experimenters and industrialists alike, we shall

Principles of diamond synthesis 17

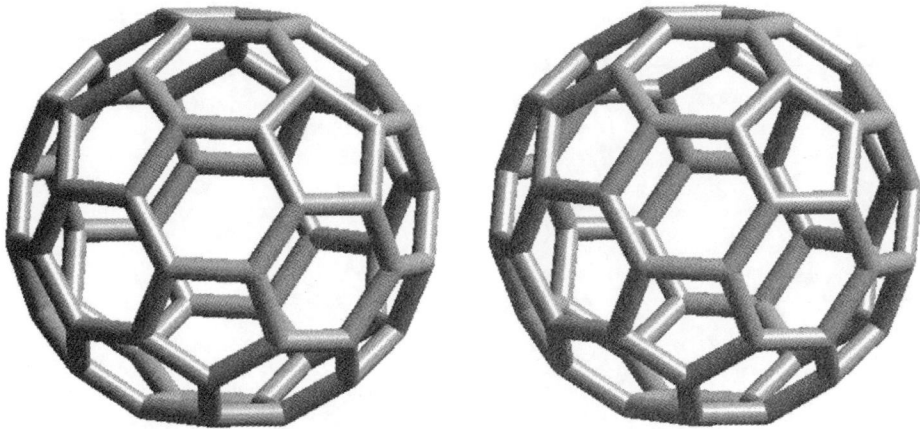

Figure 3.1 Buckminsterfullerene top (left) and front (right) view.

focus on graphite to diamond conversion in this text, beginning with the differences in the atomic structure of the two forms and how these differences create problems for synthesis.

Graphite and diamond

As so far discussed, diamond and graphite are both composed of the same element; however, the physical structures of the two siblings are exceedingly different. While

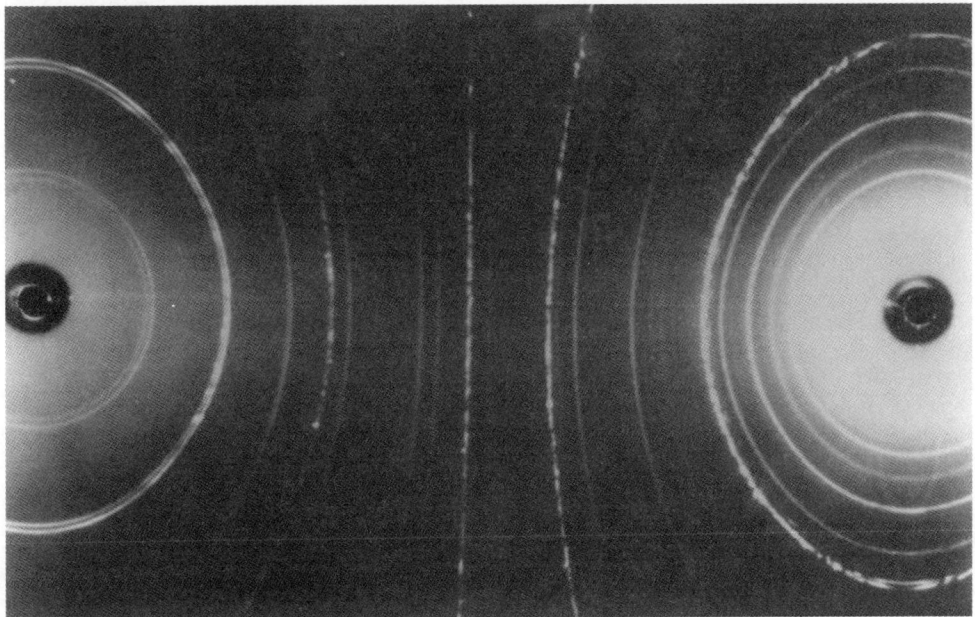

Figure 3.2 X-ray diffraction photograph for diamond, used as verification of synthesis of the first synthetic diamonds. (Courtesy H. Tracy Hall Foundation © 1998.)

18 The diamond formula

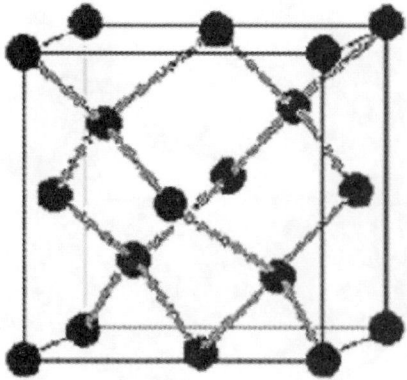

Figure 3.3 Diamond unit cell.

both forms of carbon utilize covalent bonding, the atomic lattices vary in density, orientation, crystallography and symmetry. Diamond crystallizes in the highest crystal class, the cubic system, and has the full elements of symmetry exhibited in a euhedral crystal. The unit cell incorporates fourteen carbon atoms arranged in tetrahedral patterns, however four of these are at the corners and a further six are located on faces, all of which are shared with neighbouring atoms. The result of this is that each unit cell *contains* eight carbon atoms solely dedicated to each cell. This information is known from early studies, as diamond was one of the first structures discovered by the Braggs in 1914, during their development of early X-ray crystal analysis.

Graphite crystallizes in the hexagonal system, and has reduced symmetry when compared to diamond. The structure can be described as horizontal sheets of strongly bonded atoms in a hexagonal arrangement with bonds 0.142 nm (nanometres) in length, connected via vertical bonds of 0.335 nm in length. It is the longer (and therefore weaker) bonds of the vertical direction that form the perfect cleavage, giving graphite its slippery appearance and excellent lubricating qualities.

Diamond has a much more uniform lattice, with its atomic covalent bonds measur-

Figure 3.4 Graphite atomic lattice structure with 1.42 Å lateral and 3.35 Å longitudinal bonds.

Principles of diamond synthesis 19

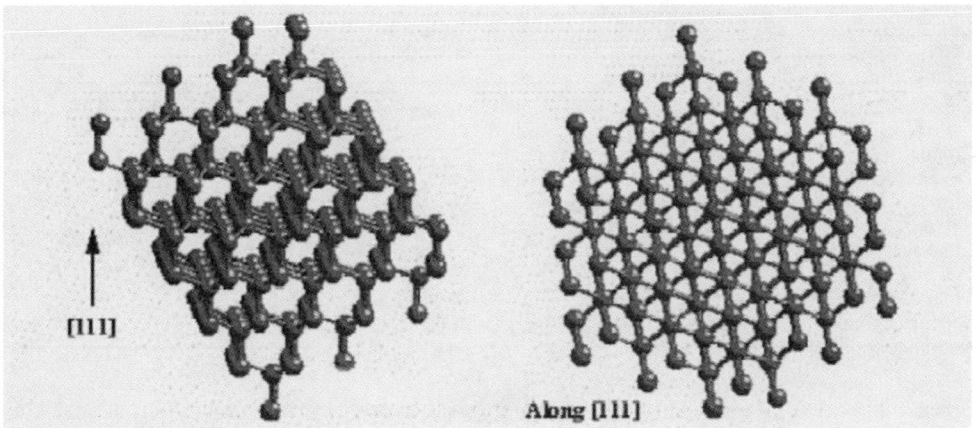

Figure 3.5 Diamond atomic lattice structure with 1.54 Å bonds in all directions.

ing 0.154 nm in all directions. While this does promote a level of uniformity in structural tenacity, diamond still possesses perfect cleavage, manifesting in the octahedral direction. This is very difficult to visualize in two dimensions as depicted in Figure 3.5. However those people who are able to construct a three dimensional model will find on rotating the structure that, in certain directions (perpendicular to each other) the lattice appears thinner in density; almost akin to the graphite structure in Figure 3.4. These are the cleavage directions in the diamond lattice. It is these cleavage directions that make diamond susceptible to breakage, as any jeweller will attest.

This brings us to the subject of density. This originally posed a major stumbling point for chemists and earth scientists of the late 1700s, in the era of Tennant. How could diamond and graphite, both formed of the same atoms, vary so substantially in density? The specific gravity (or relative density when compared with the density of water) of diamond is 3.51524 (3.51506 for type II, and 3.51537 for type I),[11] and graphite is variable around 2.24. It is now known that this is due to the closely packed carbon atoms of diamond, and the comparatively loose packing of the carbon atoms in graphite.

Out of this differential density paradox was born the concept of the application of high pressure to depress the atoms into a more closely packed arrangement. This will be discussed in the next section on pressure.

Not only is diamond and graphite different on a nanoscopic scale, but the natural habit and physical properties of diamond and graphite are also in major contrast. Diamond is (in its purest form) a transparent, colourless mineral that (as we have discussed) crystallizes in the cubic or isometric system. It forms (in perfect form) equant crystals, either octahedral, cubic, sometimes dodecahedral or twinned crystals called macles. (Diamond can also form as a micro-crystalline mass, called carbonado.[6]) It has a hardness of 10 on the Mohs scale (a non-linear scale used to rate the scratch hardness of known minerals), 140 times harder that corundum which is 9 on the same scale. The indentation hardness of diamond (on the Knoop scale) has a value of 5500 to 8500, compared again with corundum at only 2000.[12] Diamond has a nil to white streak, adamantine lustre, conchoidal fracture and perfect octahedral cleavage as discussed.[13] Graphite forms as flat, tabular hexagonal plates, or more often as a massive foliated mass. It is opaque, grey to black and with a metallic lustre and a dark grey to black

20 The diamond formula

Table 3.1 Physical properties of diamond and graphite.

Physical property	Diamond	Graphite
Common habit	Equant: octahedral, cubic, dodecahedral or twinned	Massive foliated or tabular hexagonal plates
Colour	Colourless when pure	Grey to black
Transparency	Transparent when pure	Opaque
Lustre	Adamantine	Metallic
Streak	None to white	Grey to black
Cleavage	Perfect octahedral	Perfect basal
Fracture	Conchoidal	Uneven
Mohs hardness	10	1–2
Specific gravity	3.52	2.24

streak. The cleavage of graphite is perfect basal cleavage, rather than octahedral, the fracture uneven, and the hardness of graphite (when measured on the Mohs scale) is between 1 and 2.[2]

As is evident from Table 3.1, diamond and graphite differ in all physical properties. Physical properties are not the only characteristics in which the two minerals differ greatly. Optical properties are also vastly different for diamond and graphite.

Diamond has a refractive index (the ratio of the velocity of light in a vacuum, to the velocity of light in diamond) of 2.417 (for sodium light). It is singly refractive (does not plane polarize light on entering the diamond) and has a high dispersion of 0.044 (see Chapter 8). Diamond is also affected by ultraviolet light with almost predictable results. Diamond is optically transparent to radiation with wavelengths from 2.5 µm in the infrared to 225 nm (type II) and 340 nm (type I) in the ultraviolet.[15]

Graphite has a refractive index that cannot be measured due to its opaque nature and metallic lustre, but as it crystallizes in the hexagonal system is known to be doubly refractive with a uniaxial optical sign (the refractive index is variable in only one of the two refracted–polarized rays). Graphite is not affected by ultraviolet light. The dispersion of graphite is nil.[7]

These characteristics are outlined in Table 3.2.

It is understandable why past scientists compared the synthesis of diamonds with the alchemists' attempts at making gold, when collating all of the differences as has been done. How do you turn the ultimate lubricant into the ultimate abrasive? It may appear as though diamond and graphite have nothing in common at all, with the exception of their chemical kinship. It is this chemical relationship that forms the basis for the similarities between diamond and graphite, and it is these similarities that open the door to modern science, allowing the next generation of diamond makers to enter a new era.

Table 3.2 Optical properties of diamond and graphite.

Optical property	Diamond	Graphite
Refractive index	2.417	N/A
Optical sign	Isometric	Uniaxial
Dispersion	0.044	Nil
Ultraviolet reaction	Stimulated by ultraviolet light	Inert to ultraviolet light

The similarities begin with the obvious carbon composition and covalent bonding, and from this we find other cognate properties. Both minerals are chemically inactive and insoluble in liquids, but burn to carbon dioxide when heated in oxygen.[3] Additionally (at pressures designed to avoid the graphitization of diamond) both diamond and graphite have a latent heat of fusion at 3550 to 3840°C, at latent heat of sublimation of 4200 to 4827°C and at triple point at approximately 5000K.[15] The fact that diamond and graphite share the same critical temperature and boiling point (latent heat of fusion) indicate that they can both be broken down, and more importantly, at around the same temperature or energy level (dependent upon pressure).

It is the remaining elements of diamond synthesis, high pressure and high temperature that constitute this energy and required conditions.

High pressure

To a certain degree the world of high pressure is transient. It is continually changing, mutating and its boundaries are stretching. Rather than attempting to keep pace with the advances in this area we will try to confine our synopses to the known information relating to diamond synthesis.

Pressure is defined as a force per unit area exerted at right angles to a surface (i.e. perpendicular to it). The international system of units (SI units) used in scientific work, expresses pressure in terms of pascals,[2] a measure of newtons per square metre (N/m^2). Former terms still used in non-scientific contexts often refer to pressure as kilograms weight per square centimetre, or pounds weight per square inch.

When discussing very high pressures, for the purposes of simplicity, another unit of pressure is frequently used. The term applied is 'atmospheres'. An atmosphere is defined as equal to the pressure exerted by a column of the liquid metal mercury exactly 760 mm (29.9 inches) high.[16] This unit is very close to the average pressure of the earth's atmosphere at sea level. It corresponds to 101.325 kilopascals (kPa).

Another widely used term is the 'bar', which is often used interchangeably with atmosphere, although one bar is equal to 0.987 of an atmosphere. To give some examples of atmospheric pressure levels that effect us in our daily lives, the pressure within a car tyre is approximately two atmospheres. The pressure cookers we use in our kitchens generate modest pressures of approximately half an atmosphere, while a scuba diver at a depth of around 30 feet will experience almost ten atmospheres.[2]

The pressures we experience are very modest when compared with the high pressures employed by pressure physicists in the laboratory, or the ultra high pressures induced by the earth's great mass below the mantle due to its gravitational acceleration. Why has pressure been of interest to science? Under pressure, matter undergoes many amazing transformations, so much so that a whole discipline of physics is devoted to quantifying and qualifying these changes and the resulting phenomena.

Many great scientists decorate the history of high-pressure research including Percy Bridgman, Loring Coes, the eccentric Baltzar von Platen, and the controversial George Kennedy, all of whom play a part in this story. Perhaps some of the greatest research (and discoveries) have come from the collaboration of teams of experimental researchers such as the General Electric team (the subject of most of Chapter 5) and the US National Bureau of Standards diamond cell inventors Alvin Van Valkenburg and Charles Weir. Many excellent texts are available for those interested in high-pressure work, but in the interests of contextual diversity, ours is a brief description.

Current pressure physics tells us that almost all matter has a number of states that are stable at different pressures. The achievements of Bridgman and Coes (after whom the high-pressure form of quartz, Coesite, is named)[2] alone account for the synthesis of hundreds of high-pressure minerals, including the before mentioned five solid forms of water discovered by Bridgman in 1911. As we are primarily concerned with diamond synthesis here, we will use carbon as the basis for our discussion on the effects of high pressures.

At normal or low pressures, graphite is the solid state favoured by carbon (and the first to form, given the opportunity); however, at higher pressures such as those deep within the earth, diamond is the stable state and will form more readily than graphite. Synthetic diamonds are formed in the laboratory in pressure conditions very similar to the pressures experienced by natural diamonds within the earth. These conditions do vary, but are usually around 55 000 atmospheres,[1] to 100 000 atmospheres.[2]

Scientific theorists predict that at around 12 megabars (11.84 million atmospheres) diamond undergoes another phase transition,[17] becoming a new material with as yet unknown properties. This material, named SC4, has been simulated on super-computers and is believed to be metallic in appearance. It is also presumed to be metastable at up to about 3 terapascals (29.6 million atmospheres) where the diamond lattice collapses.[18] Additionally another form is believed to exist around 7.8 megabars (7.7 million atmospheres) which is considerably less than the predicted 11.8 Mbar (11.6 million atmospheres). This form has been called BC8.[19]

To duplicate the pressures deep within the earth, techniques and apparatus have been developed with specialized needs in mind. The apparatus designed and used to create diamonds is not capable of attaining the ultra high pressures of the megabar, and conversely the apparatus used to break the megabar (such as the diamond anvil cell) is inappropriate for use in diamond synthesis due to the microscopically small size of the samples.

The high pressures used during diamond synthesis are obtained by use of large presses employing the same principles adopted by Bridgman with his tapered anvils. The multiplication of pressure is achieved by the strong contrast in the ratio of the surface area of the interface with the sample and the surface area contacting with the originating force. The example for this is supplied in Chapter 2.

By exposing a carbon sample to these pressures the atoms comprising the sample are encouraged to align themselves and bond in the configuration most stable at that particular pressure. At low or normal pressures that stable form is graphite. At the higher pressures outlined above (10 000+ atmospheres) that form is diamond.[2]

But what if the carbon atoms have already aligned themselves? How can we change this trend? Bridgman could attribute to the problems this posed. In all of his experiments the graphite sample always reverted to its original hexagonal form when the pressure was removed. The solution to this enigma lies in the third element of diamond synthesis, temperature.

High temperature

Unlike high pressure, we can see the transforming effects of temperature every day. The temperamental weather conditions caused by atmospheric temperature fluctuations, the evaporation of steam from a boiling kettle, or the melting of an ice-cream. Why does temperature have this effect on matter – the effect of changing matter from one state to another? Before delving into the power of temperature, we must first understand what temperature is, and how we are able to measure its power.

Temperature is defined as a property of systems that determines whether they are in thermal equilibrium,[20] so the temperature of a system can be considered to be a measure of the energy of the atoms and molecules composing the system. In this example we are referring to kinetic energy due to their motion. Temperature scales are our way of relating this energy to the thermal energies of known quantities. One of the earliest temperature scales, devised by the German physicist Gabriel Daniel Fahrenheit, measured temperature against the transition points of water at standard atmospheric pressure. His freezing point (and melting point) of ice is 32°F, and the boiling point of water is 212°F.

Throughout most of the world the centigrade, or Celsius scale, invented by the Swedish astronomer Anders Celsius, is used rather than the Fahrenheit scale. This scale assigns a value of 0°C to the freezing point of ice and 100°C to the boiling point of water.

As is the case with many every-day measuring techniques and scales, science required a more precise and accurate system on which to calculate theories and to test them. The British mathematician and physicist William Thomson (1st Baron Kelvin), devised the absolute or 'Kelvin' scale, which is in fact a more natural scale. In this scale, absolute zero is at − 273.15°C, which is 0K. This is the state when the mean kinetic energy is zero. The degree intervals are identical to those measured on the centigrade scale. The corresponding 'absolute Fahrenheit' or Rankine scale, devised by the British engineer and physicist William J.M. Rankine places absolute zero at − −459.67°F, which is 0°R, and the freezing point at 491.67°R.[21] We will be referring to the Celsius (or centigrade) scale in this text, unless otherwise stated.

We have learned to think of temperature in terms of a comparative scale with which to measure the hotness or coldness of a given medium. An increase in heat will be expressed as a higher level on the scale, therefore when we make reference to temperature we are normally referring to heat, or a quantity of heat.

Briefly, heat is actually one form of energy. Energy is measured scientifically in joules (J). For heat to be produced for us to measure with our wide range of temperature scales, energy must first be transferred, or applied and then exhausted releasing heat as a by-product (exothermic). An example of this is the way electronic devices tend to become hot to the touch after they have been in use for some time. So heat is then described as the transfer of thermal energy from one system to another, as a result of a temperature difference arising in the systems. Therefore the process of heating a substance is the same as imparting energy to it. Physics tells us that the energy may be in the form of electric potential, electromagnetic radiation, or heat transferred from external (to the system) sources, for example. The energy, in whatever form, excites the atoms and stimulates their electrons temporarily into an excited state (higher energy level). As the electrons later return to their original positions energy is released, possibly in the form of heat.[20]

Imparting heat to a substance induces a phase transition, because as the electrons become excited their bonds with surrounding atoms are broken. A phase transition is the change of a substance from one form to another, such as a solid to a liquid, or a liquid to a gas. The required level of energy needed to break atomic bonds is unique to a particular substance and is called the 'latent heat of fusion' for that particular substance. If water is heated the hydrogen and oxygen molecules release the atomic bonds with surrounding molecules and at first bubbles of gas (water vapour) form, rise to the surface and dissipate into the atmosphere. Heat the water still further and more atoms become excited; the result is steam. The water droplets are the remaining tiny groups of atoms still clinging together. Water cannot be transformed into steam without stimulation from an independent energy source.[21]

The same applies to other forms of matter. Both diamond and graphite melt at above 3550 to 3840°C and revert to gas at around 5000°C. This equates to the level of energy applied (in the form of heat) to the material to enable the atoms to release their covalent bonds, and transform into the corresponding phase.

It is possible to control, and effectively harness the transfer of heat because, as shown by the laws of thermodynamics, heat always travels from the hottest substance to the cooler in order to achieve equilibrium.[22, 23] Another factor in the control of heat transfer is the ability of some substances to attract or conduct heat far more efficiently than other substances. These materials are called thermal conductors, and both diamond and graphite fall into this category.

The only method of heat transfer in opaque solids is conduction. If the temperature at one end of a metal rod is raised (by heating), heat is conducted along the rod to the colder end. The exact mechanism of heat conduction in solids is not entirely understood, but it is believed to be partially due to the motion of free electrons, which transport energy if a temperature difference is applied.[22, 23] This theory helps to explain why good electrical conductors also tend to be good heat conductors. Carbon is a good conductor of heat; while not as efficient as copper and silver in *every* situation for example, it has its uses in industry as an efficient absorber of heat. Diamond especially forms excellent heat sinks, as it has the highest thermal conductivity of any crystalline material.[24] As will be seen in Chapter 4 the thermal conductivity of carbon plays an important role in the synthesis of diamonds.

Another property of heat that plays a part in the synthesis of diamonds is similar to, but not the same as, conduction. If a temperature difference arises within a fluid (liquid or gas), then fluid motion will almost certainly occur. This transfers heat through the same thermodynamic process already discussed, from one part of the fluid to another. This process is called convection. If a liquid or gas is heated, its density (mass per unit volume) generally decreases due to expansion. If the liquid or gas is in a gravitational field as we are on earth, then the hotter (less dense fluid) rises allowing the colder, denser fluid to sink in accordance with the gravitational acceleration. This kind of motion, due solely to the variation in the fluid's temperature, is what constitutes the kind of natural convection used in diamond synthesis.

An every-day example is that of water in a pan being heated from below by a hotplate or flame. The liquid closest to the bottom is warmed by heat conducted through the bottom of the pan (thermal conduction). It expands and its density decreases, as a result the hot water rises to the top and some of the cooler fluid descends towards the bottom. This forms the beginning of a circulatory motion. The cooler liquid is again heated by conduction, the warmer liquid at the top loses its heat by conduction and radiation into the air at the top of the pan.

Convection can apply both to gases as well as liquids. Heating a room using a small electrical heater depends on natural convection currents to heat the whole room. The hot air rises along the wall, displacing the cooler air and sending it back down toward the heater, where the process continues. Convection also determines the movement of large air masses above the earth, the action of the winds, the formation of clouds, ocean currents, and the transfer of heat from the interior of the sun to its surface.

Now we have a basic understanding of the major elements of diamond synthesis we will examine how these ingredients are put together, before we continue our journey down the road toward success.

Chapter 4

Putting the pieces together

To understand how diamonds could be created in the laboratory, scientists turned their attention to analysing the conditions under which natural diamond formed. Such scientists already had a working knowledge of the emplacement process through lessons learned with mining and exploration; however, this was only part of the overall synopsis to be compiled. Formation of natural diamonds is a distinct process from emplacement, and unravelling the secrets from deep within the earth is a daunting task.

Genesis of natural diamonds

It is now known that natural diamonds crystallized from ultramafic magmas from around 3300 million years ago (Kimberly, South Africa) to 990 million years ago (Orapa, Botswana). The two distinct types for rocks in which natural diamond crystallized are termed peridotite, a course-grained ultramafic rock composed primarily of olivine with additional minerals such as pyroxenes; and eclogite, a course-grained rock composed primarily of almandine-pyrope garnet and clinopyroxene.[10] It is also known that natural diamonds are found as zenocrysts, in transporter extrusive rocks called kimberlite and lamproite, and that these rocks are purely transporters playing no part in crystallization. Study of zenoliths containing natural diamonds, and the inclusions in natural diamonds have assisted in drawing conclusions about the pressure-temperature regime in which natural diamonds crystallize.[10]

In the past the study of natural diamond formation fuelled the research on diamond synthesis, however times have changed. Logically one would deduce that by reproducing the conditions under which natural diamonds crystallize one could reproduce synthetically what was once exclusively a product of nature. Ironically today the flow of information is to some degree reversed. Known facts relating to the synthetically produced high-pressure minerals (subsequently found to exist in nature) have shed some light on the origins of their natural counterparts.

An example of this is the mineral coesite. Coesite was first synthesized by Loring Coes, in the laboratories of Norton in 1953. It is a high pressure, dense form of silicon dioxide formed at 35 000 atmospheres and 800°C, conditions equating to depths of less than 50 miles below the surface of the earth.[2] Coesite was for a long time thought to be only a synthetic mineral, but was later found at the famous Canon Diablo Meteor site.[2] This is significant as coesite also forms as an inclusion in natural diamonds. The known crystallization conditions for coesite (and other high-pressure minerals found in diamonds) send an important message on the conditions under which the host mineral (the natural diamond) formed.

Inclusions in peridotitic natural diamonds (p-type) crystallized at temperatures ranging from 900°C to 1300°C, and pressures ranging from 45 000 to 60 000 atmospheres. The estimated depths for formation of natural p-type diamonds are in the range of 150–120 km below the surface, in the earth's mantle. Inclusions found in eclogitic natural diamonds (e-type) suggest a greater crystallization temperature, and form at greater depths than p-type natural diamonds.[10]

Replicating nature

To use a familiar analogy, carbon, high pressure and high temperature (heat) provide the ingredients for a kind of 'synthetic diamond formula'. In diamond synthesis, as with all good formulas, there is a correct sequence and appropriate measures. It is these measures and sequences that proved problematic for scientists including Bridgman and his predecessors. There is a need for the right combination, but why is it so important?

The main oversight of many early would-be diamonds makers was the omission of the heat factor. Pressure alone is insufficient to convert graphite to diamond; as Bridgman discovered, it will revert to its original form when the pressure is removed. Additionally heat alone will merely melt the graphite, only to have it re-crystallize as graphite upon cooling. The two must be combined to give the right result.

The omission of the element of heat in the process principally removed a major part of the operation. As we have discussed, applying heat to a substance excites the electrons of the atoms comprising the substance, and encourages them to release their atomic bonds. This is the key that was missing in many early attempts. Pressure alone succeeds only in squashing the original substance and distorting its atomic lattice, but as the bonds are distorted (and not actually broken) the atoms cannot change their configuration. Remember that each carbon atom in graphite has three bonded neighbours, and carbon atoms in the tetrahedral diamond lattice have four neighbours.[2]

The addition of heat in the case of graphite to diamond conversion is responsible for breaking the covalent bonds of the graphite configuration, leaving them free to re-align in a new arrangement if the conditions permit. Once 'separated' the carbon atoms seek to re-establish new bonds as they are allowed to cool. Depending upon the pressure conditions the new form may be graphite, or it may be diamond. Pressure is the component that provides the appropriate conditions. At normal pressures graphite is the thermodynamically favoured crystalline state for carbon.[6] That means, under normal conditions, carbon will crystallize as graphite because this is the state requiring the least kinetic energy or work. It forms the graphite lattice arrangement first.

At high pressures this is not the case. Due to the pressures acting against the carbon atoms, the state of requiring least kinetic energy is a more dense, more closely packed arrangement. Diamond is the thermodynamically favoured state under these conditions, and therefore the carbon atoms readily align themselves in the tetrahedral lattice of diamond.[6] At high pressures, graphite will not be able to form, as this will take more work than if diamond were formed, just as diamond will not form at normal pressures. Based on this scenario, carbon heated (to the point of the elimination of the graphite lattice bonds) under high-pressure conditions will re-crystallize as diamond upon cooling; and this is just what happens.

There is also a very close relationship between the temperature and pressures required to stimulate conversion. In 1938 Frederick D. Rossini and Ralph S. Jessup constructed the first series of ratios or co-ordinates that indicate at just which pres-

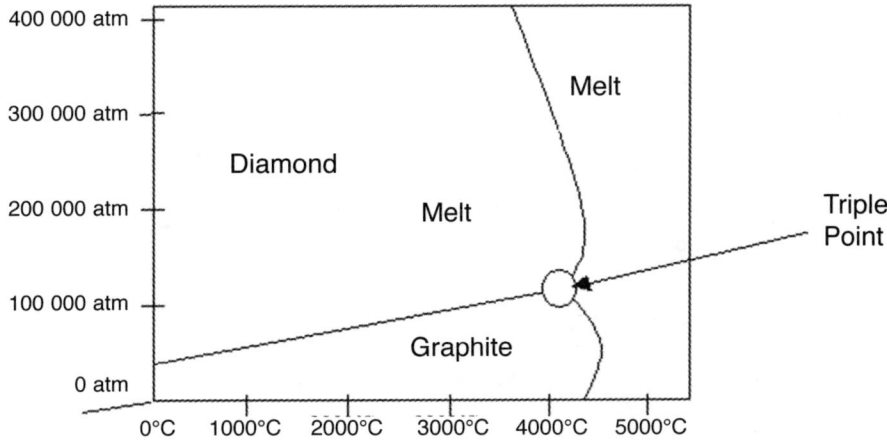

Figure 4.1 Phase diagram for carbon including triple point and corresponding pressure (atmospheres) and temperature (degrees centigrade) conditions.

sures and corresponding temperatures conversion will take place. This is called a 'phase diagram'.[1]

Above the conversion line in the phase diagram in Figure 4.1 are the conditions under which diamond is the thermodynamically favoured state; and under the line graphite is the thermodynamically favoured state. The conversion line is also known as the Berman-Simon line, after the scientists R. Berman and F. Simon, who in 1955 extended and refined the former calculations.[1]

Above the line is the thermodynamically stable region for diamond, and to the right the area indicates the point at which both diamond and graphite melt. Notice that there is a point at which graphite, diamond and a molten concoction of carbon can co-exist. This is called the triple point, has been approximated at 5000K and 12 Gpa.[144]

After the atomic bonds of the graphite are broken, the conditions must remain in the region above the line to induce diamond to form; that is, the most consistent and uniform conditions of synthesis correspond to those indicated above the line on this diagram.

This is where sequencing becomes important. If pressure is removed or reduced before temperature, the carbon atoms may not have re-aligned to form new bonds. Reducing the pressure may change the conditions to those corresponding to the area under the conversion line in our phase diagram, thus producing graphite once again.

The temperature must be shut off and the sample allowed to re-solidify, before the pressure is removed to ensure that the diamond lattice has formed. Once formed, the diamond lattice is a solid, and is stable at room pressures. It will not suddenly revert to graphite upon removing it from the device.[6]

The temperature/pressure ratios effectual (and cost effective) in the synthesis of diamonds range from 1000°C to 2000°C, and 50 000 atmospheres to 100 000 atmospheres. As depicted in the phase diagram, the higher the temperature employed, the higher the pressure required.

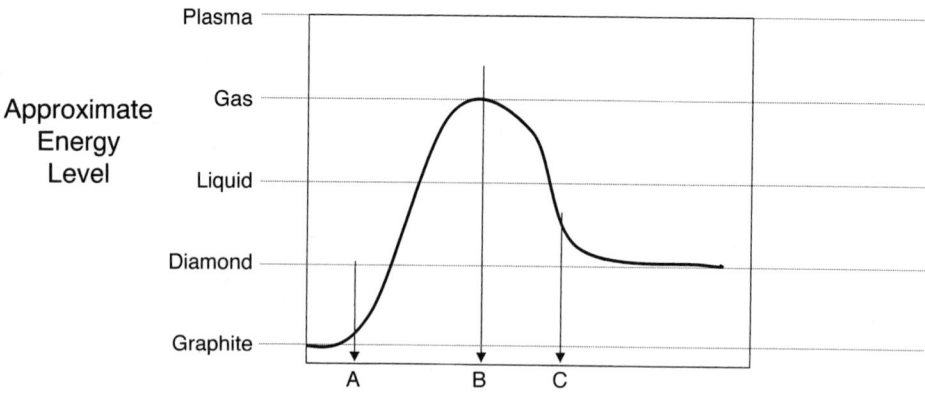

Figure 4.2 Conversion diagram for graphite to diamond conversion with starting conditions (A), activation conditions (B) and reaction conditions (C).

Figure 4.2 is a graphical representation of the sequence of the procedure.[25] To represent the levels of energy (in the form of heat) required to stimulate conversion at different points during the process, we have employed four of the recognized states of matter.[4] As we are referring in this case to carbon, we have added graphite because this is our starting material. We have used the graphite phase to represent the lowest energy phase, followed by diamond, liquid, gas, and plasma, being the highest energy phase of matter[26] (see Chapter 14 for further discussion on the role of plasmas in diamond synthesis).

Assuming that this experiment is performed in a high-pressure device, at around 70 000 atmospheres, the following is a basic description (aided by reference to Figure 4.2) of what may occur. The energy is first applied at point A (indicated on the diagram), exciting the atoms and effectively elevating the carbon into a higher energy level (approximating that of a gas). Here all the atomic bonds are effectively broken to the degree required by the experiment. At point B the energy is reduced (or shut off) and the carbon atoms re-arrange themselves in the most thermodynamically favoured state, in this case diamond, at point C.

It is important to note that the energy required to stimulate this conversion is much greater than merely elevating the energy level straight to that of diamond. It is almost as if an obstacle must be overcome before settling in the appropriate position.

There are other series of concerns that play a part in the synthesis of diamonds, economic concerns. To maintain high temperatures and/or high-pressures for extended periods of time is expensive. The lower the temperature or pressure the lower the cost, so the actual ratio for conversion used is situated just above the conversion line on our phase diagram (Figure 4.1). This represents the lowest possible ratio of temperature and pressure that will yield synthetic diamonds. This remains a very expensive procedure.

Catalysts

To tackle these economic concerns, and to enable more control over the process, catalysts were introduced. Catalysts are metallic solvents and have been the subject of intensive experimentation and development in their own right. The catalysts act as a kind of flux and play a number of roles in the procedure.

First, the presence of a catalyst metal in the reaction chamber with the carbonaceous material assists in ensuring that the hexagonal atomic bonds of the graphite are broken. The metals are excellent conductors of heat and melt evenly at a pre-determined temperature, encouraging the non-diamond carbon to dissolve. As carbon is a thermal conductor the graphite will readily transfer heat from these catalytic solvents. This translates to a lower temperature level required to achieve uniform melting and breaking of the bonds, making for more efficient and cost effective production.

By using the same graphical representation as ascribed to the previous description of conversion, we can see what type of impact the catalyst has on the required energy levels.

The graph in Figure 4.3 ascribes the same importance to the points A, B, and C. The dotted line represents the energy levels attained in an experiment without a catalyst, as in the previous section. The solid line represents the energy levels for an experiment that includes a catalyst.[25] Note that the carbon atoms achieve the required energy level to break the atomic bonds together with the metal catalyst, at around the level equating to that of a liquid. This represents a significantly lower energy level required to achieve the results, and this in turn represents a cost saving.

Second, research into the effectiveness of different metals has concluded that not only are certain metals better suited as catalysts, but that different catalyst metals require specific temperature/pressure conditions to perform efficiently. The metals most frequently used as catalysts are iron, nickel and cobalt,[1] as well as manganese, palladium, platinum and alloys of these metals.[6] Information relating to alterations to the chemistry of catalysts for specialized purposes is given in Chapter 7.

Using nickel as an example, it is easy to illustrate the accuracy that a particular catalyst metal offers. The required conditions can be predicted as approximately 1500°C and 90 000 atmospheres for the purposes of this example. The nickel dissolves about

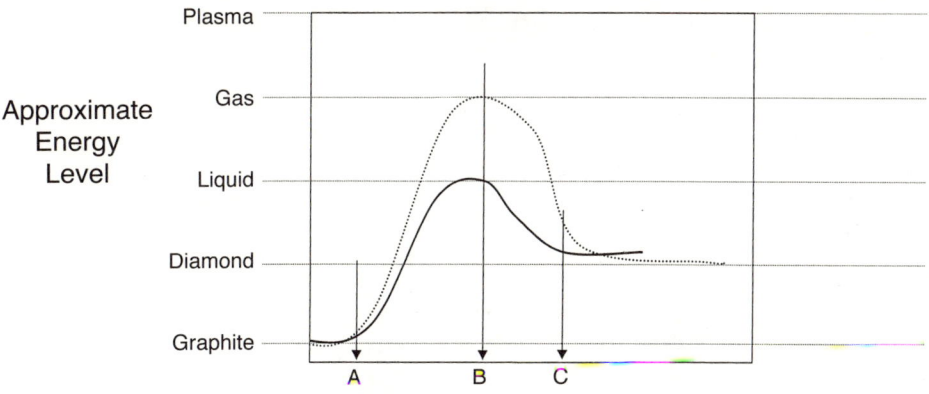

Figure 4.3 Conversion diagram for graphite to diamond conversion with catalyst (solid line) and without catalyst (dotted line), including starting conditions (A), activation conditions (B) and reaction conditions (C).

4 per cent of the carbon from the graphite under conditions of graphite stability. However under the conditions of diamond stability only 3.6 per cent will be dissolved, with (subject to the pressure conditions) the difference of 0.4 per cent being crystallized out as diamond.[6]

Catalysts such as palladium and platinum allow the use of temperatures of around 2050°C to 2500°C, much higher than for other metals.[27] This can provide additional benefits during production, including the advantage of being able to increase the variety of carbonaceous materials. For example charcoal or naphthalene can be used as source materials when employing these particular catalysts.

In effect the catalyst metal acts as a type of solvent,[1] and should be considered the fourth element in the diamond synthesis formula. The exact role of the catalyst metal is still a point of debate among different schools of thought, however the importance is undeniable.

While in hindsight it is relatively easy to isolate the conditions under which diamonds are synthesized, initially the discovery came slowly, following a concerted effort by some of the world's major development companies. The description here constitutes a basic overview of the processes involved in diamond synthesis and the important constituents. This knowledge with now be related to the international race toward eventual success, and the major players in diamond synthesis research of the time.

Chapter 5

Success in diamond synthesis

Following World War II many companies felt an increased urgency to synthesize diamonds. This was mainly due to the shortage of industrial diamonds required for machine equipment and munitions during the war years, and to the promise of great wealth. Many of the projects were bound in secrecy. Researchers were forbidden to publish successes, no matter how large or small. The companies vying for the ultimate prize spanned the globe, all pursuing the same goal but working (totally) independently, without realizing just how close the race actually was.

There has been much conjecture over who was actually first to synthesize diamonds in the laboratory. The American General Electric Company (or GE) was first to patent their process and release their results, however the Swedish company Allmanna Svenska Elektrika Aktiebolaget, or ASEA (which translates to Swedish General Electric Company) claimed for many years their team achieved synthesis of diamonds well before the GE team. Before we discuss the GE process and patents we will briefly look at the ASEA results.

Allmanna Svenska Elektrika Aktiebolaget

Allmanna Svenska Elektrika Aktiebolaget, or ASEA, began their quest for synthetic diamonds in the early 1940s. The device they constructed for the purpose was by all accounts a large, impractical, very expensive apparatus designed by the eccentric Baltzar von Platen, who joined the project in 1942. Von Platen's greatest invention was the portable thermal refrigerator that produced ice with a gas flame,[2] which was an ingenious invention at the time. Given his success in his previous ventures he was considered to be the most highly qualified man for the job.

Von Platen used his unique mind to devise his own machine for the purpose of making diamonds. His approach to the design of the high-pressure apparatus was, in a way, back to front. To his way of thinking it was inevitable that the device would be destroyed in order to achieve the desired pressures and temperatures. In order to save time, and expense, his device was 'already broken'.

The design to be constructed was a high-pressure chamber formed by a number of independent components that served a specified purpose. Each was isolated from the others, so that when some parts were damaged the remaining pieces were still operational. Additionally, it was intended that this unconventional approach would reduce the problem of the high temperatures rupturing high-pressure devices.[2]

Von Platen agreed to provide designs for an innovative machine, including any high-pressure equipment he already possessed, and the ASEA engineers constructed the remaining components. Von Platen advised the engineers during construction;

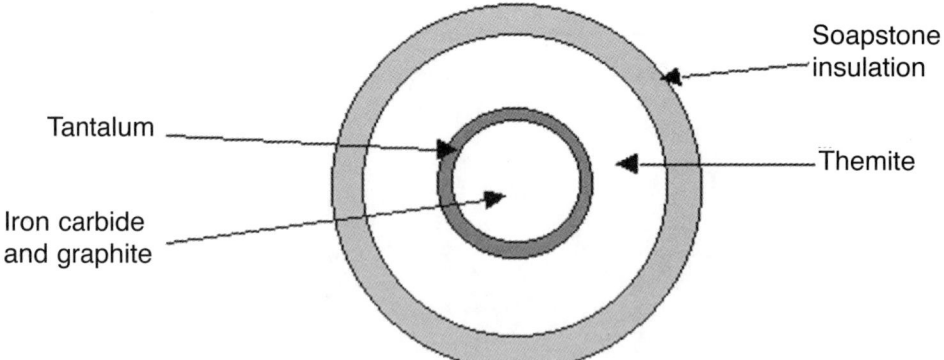

Figure 5.1 Schematic diagram of internal component of the ASEA split sphere high-pressure device.

however, upon completion, his role shifted and the project was left to the ASEA team, named QUINTUS. The device resulting from months of preparation was the split sphere apparatus (Figure 5.1), designed to achieve the same uniform pressure for all sides of the sample, approximating the conditions in nature.

It comprised a sample about the size of a marble, surrounded by a sphere of thermite about the size of a fist. Thermite is a compound of barium peroxide and magnesium that burns at 4000°C. The sphere was then contained within a soapstone shell, inside a cube-shaped copper or iron jacket. This reaction cell was placed inside an awkward chamber formed by the correct arrangement of six steel and carbide, wedge-shaped anvils.[28]

The anvils alone however, are not sufficient to apply the necessary pressures without a driving force. The entire split sphere device has to be pressurized; and most importantly, evenly from all directions. To do this the half-tonne, six-anvil sphere is placed in a water-tight copper cover and then placed inside a cylindrical chamber containing water, that was then pressurized to around 6000 atmospheres.

The concept was designed around the water pressure experienced by a diver, though the tank pressure was much greater. The problem of the chamber bursting under the pressure of the water, or the water leaking, was overcome by wrapping the entire chamber in coil upon coil of piano wire, and placing it inside a welded, re-enforced steel yoke. Even with the layers of iron and steel the device still became fatigued and components had to be replaced. It was anticipated that the apparatus would achieve pressures around 50 000 atmospheres and temperatures for around 4000°C by igniting the thermite via an electrical charge.[28]

Assembling the massive machine was labour intensive and time consuming. The project saw a number of set backs and even had to relocate in 1949. In 1951, with still no evidence of success, the sample was changed from pure graphite to a mixture of carbon and metals, and tested in 1952. During the same year, von Platen left the project entirely. In 1953 their tack changed again. Instead they used a mixture of graphite, iron carbide and a small amount of diamond powder. Then, on 16 February 1953, on completing an hour-long experiment estimated to have been approximately 83 000 atmospheres, the tiresome chore of analysing the sample, and cleaning it in acid, yielded tiny crystals no larger than a grain of sand. Diamonds.[2]

This was their finest hour. The crystals had not been in the original mixture, and were subsequently independently verified as being diamonds by Professor Arne

Ölander, dean of the Chemical Faculty at the Institute of Inorganic Chemistry, University of Stockholm (and member of the Royal Swedish Academy of Sciences), and his assistants, using X-ray analysis. The ASEA team had succeeded. The experiment was later repeated and more diamonds were produced. Unfortunately the process was costly and impractical for commercial use, and while the QUINTUS team struggled to fine-tune their technique,[29] the American team at General Electric beat them to the finish line.

Had they published and patented their method, history would record ASEA as the first company to synthesize diamond. Many theories circulate as to exactly why the team waited, through Erik G. Lundblad (Vice President of ASEA and former member of the diamond development team) stated in a letter published in the *Journal of Gemmology* (April 1986), that in accordance with company policy at the time, the team kept their achievements under very tight security until the process could be fully understood, and the commercial prospects maximized. ASEA had concluded that only the apparatus could be patented, rather than the process, and did not realize until seven years later that this was a misconception. Such discussion may now be considered merely academic, as the GE team are remembered as the first to succeed.

General Electric Company

Following the lead of the joint venture with Norton (dissolved in 1950), GE formed a specialized team in 1951 with the sole intention of being the first to synthesize diamonds. The project, entitled Project Superpressure, was located in the GE research laboratories in Schenectady, New York. The establishment of this project marked a very important and historic decision at General Electric; sending a message that diamond making was a serious business. General Electric had high hopes for success and poured seemingly unlimited resources into the project in its initial stages.

The founding members of Project Superpressure were drawn from the pool of scientists employed by GE. The project was managed by the engineer Anthony J. Nerad, and staffed by the physicists Francis Bundy and Herbert Strong. Initially they drew from the Bridgman research, as this was the most up-to-date and innovative perspective available at the time.[6]

In June 1951 the Project Superpressure team prepared a brief document outlining the immediate problems associated with diamond synthesis. First was the problem of designing a device capable of sustaining the high pressure and high temperature required simultaneously. Second was the need to procure the necessary equipment and resources. They began with the best press and relevant technology available to them, that of the Bridgman tapered anvils. With much difficulty they also sourced the material required for the non-extruding gasket, pyrophyllite.[2]

The first experiments involved running an electrical current directly through graphite discs, the resulting temperature around 1400°C, achieved at approximately 150 000 atmospheres.[2] These runs were unsuccessful.

The next step was to procure better equipment, and more assistance. In late 1951 technician James E. Cheney joined the team to assist Strong, along with another engineer, Harold P. Bovenkerk. Also joining the team to fill the vital role as chemist was H. Tracy Hall, who was later to instigate the breakthroughs of Project Superpressure. On 31 December 1951, the final member of the historical team, Robert H. Wentorf, then a twenty-five year old chemical engineering graduate, joined the ranks.[30]

34   The diamond formula

Figure 5.2 Press release photograph of the GE team. The press in the background is not the press that was used to make the first diamonds. (Courtesy H. Tracy Hall Foundation © 1998.)

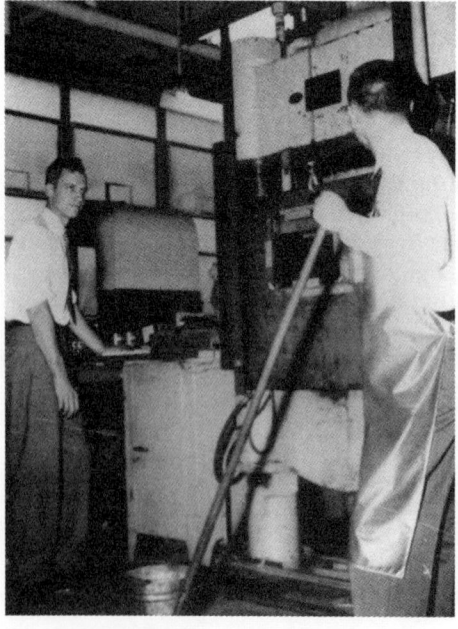

Figure 5.3 H. Tracy Hall at GE with the 'water press' (hydraulic press) used to make the first diamonds. The belt apparatus fixed within this press. (Courtesy H. Tracy Hall Foundation © 1998.)

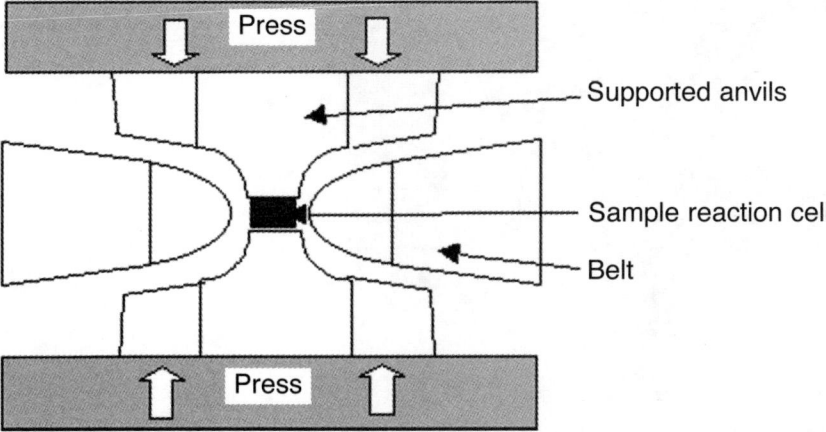

Figure 5.4 Schematic diagram of the Hall belt apparatus including reaction cell, massive support and belt.

With all of the players in place, the strategy encouraged each member to work as part of the team, but also to follow his own intuition. Nerad procured finance to purchase some natural diamonds both to study and to use as seeds. It was Bundy who devised the first of the specialized apparatus, a modification on the theme of the Bridgman anvils. The device could achieve temperatures up to 2700°C and was used throughout the experiments of 1952.[30] The only weakness of the device was its inability to achieve the required pressures, only reaching around 35 000 atmospheres, and consequently failing to create synthetic diamonds.

In early 1953 it became apparent that Bundy's device was not going to achieve the desired results and the Superpressure team set about designing new apparatus. Hall, Strong and Wentorf all designed, constructed and tested new machines devised around the main motif of a piston-in-cylinder. During the course of the year both Strong and Hall perfected their individual devices, and so the rivalry and animosity began.

Strong's device was a cone style apparatus with a unique gasket arrangement that could achieve simultaneous temperatures of 2000°C and pressures of 40 000 atmospheres. Hall's modification on the piston-in-cylinder was very different from any previous high-pressure device and was built around a half-belt design. He later updated the design to incorporate a full-belt employing two supported opposing, acutely tapered anvils and a donut-shaped ring.[2] This design was capable of approximately 4000°C and 100 000 atmospheres. The pistons and belt were formed of cobalt bonded tungsten carbide[6] and the supporting steel rings were shrunk on by heating (and thus expanding) them, slipping them on, and allowing them to contract into place upon cooling.

During the early part of 1954, tests were performed with both the cone and belt designs. Both scientists tried a variety of catalysts and carbonaceous materials, and had a few false alarms. General Electric gave the Superpressure team a deadline. On 8 December 1954 Strong seeded a graphite sample with a small diamond piece, wrapped it in iron foil, placed it into the cone device. He compressed it to around 50 000 atmospheres at 1250°C and let the experiment run over night. The following

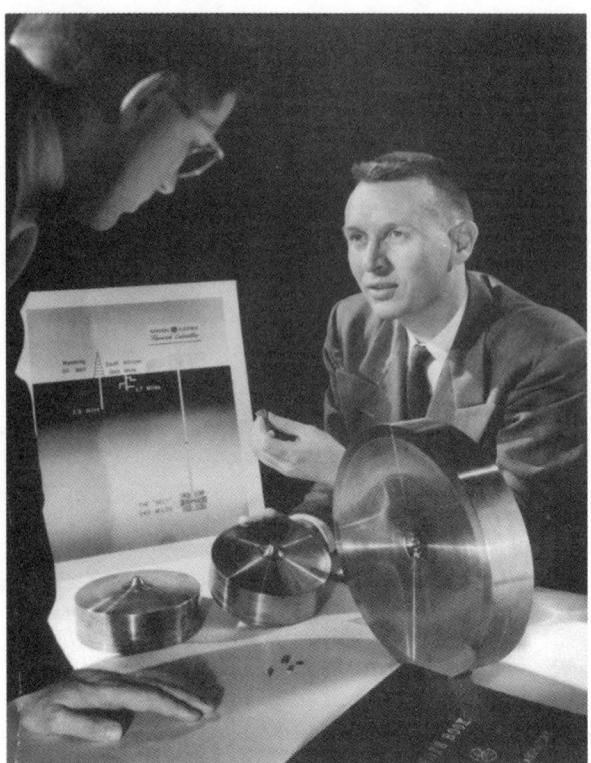

Figure 5.5 H. Tracy Hall at GE with the belt apparatus in 1954. (Courtesy H. Tracy Hall Foundation © 1998.)

Figure 5.6 H. Tracy Hall looking at synthetic diamond samples through the microscope. (Courtesy H. Tracy Hall Foundation © 1998.)

Success in diamond synthesis 37

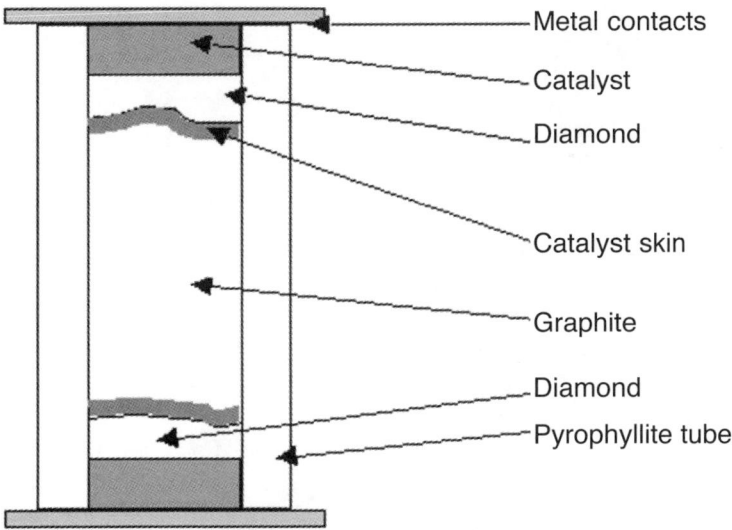

Figure 5.7 Belt device reaction cell, diamond conversion induced by an internal thermal gradient to produce industrial synthetic diamond grit.

Figure 5.8 The sample containing some of the first man-made diamonds after an early run. (Courtesy H. Tracy Hall Foundation © 1998.)

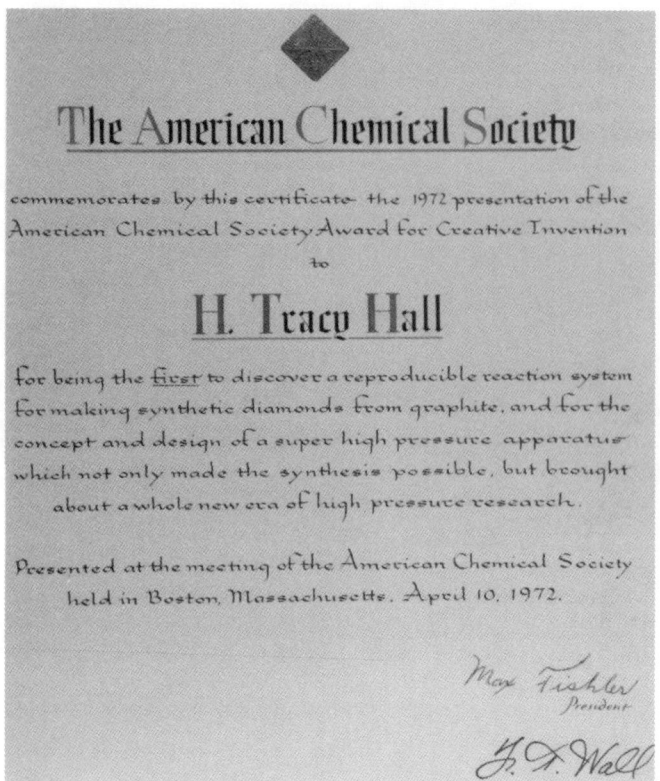

Figure 5.9 The award presented to H. Tracy Hall by the American Chemical Society, for making synthetic diamonds and opening doors in the field of high pressure. (Courtesy H. Tracy Hall Foundation © 1998.)

Figure 5.10 Close view of some of the first synthetic diamonds made by H. Tracy Hall at GE. (Courtesy H. Tracy Hall Foundation © 1998.)

day the sample was removed and upon analysis, yielded two synthetic diamond crystals, independent of the natural diamond seeds.[2,31]

On 16 December 1954 Hall performed a similar experiment using the belt apparatus. He seeded iron sulphide in a graphite heater, placed it in the reaction chamber, and exposed it to around 100 000 atmospheres and 1600°C, for a period of 38 minutes.[2] Upon removing the sample hundreds of tiny synthetic diamond crystals were discovered.

The transformation within the reaction cell was dramatic. The chemical reaction that had occurred can be described in terms of a slow convection, moving the metal catalyst through the graphite and leaving diamonds in its wake. The reaction cell is constructed from a tube of pyrophyllite, with metal contact plates on the ends (Figure 5.7). An electric current is passed through one piston to the metal end plates, heating and melting the metal catalyst and graphite sample, returning through the opposing metal plate and piston, to produce a circuit. The synthetic diamonds were then liberated, by dissolving the resulting metal and carbon block in acid.

Both GE Superpressure scientists had succeeded in producing synthetic diamond; however, to claim a scientific discovery the process must be able to be repeated, and by an independent party no less! On 31 December 1954, GE physicist Hugh Woodbury repeated the belt experiment and produced synthetic diamonds with ease. Strong could not repeat his success. Hall's device and process was patented in the United States and research began on fine-tuning the technique for commercial use.[31]

De Beers

The news of the GE success was released via a press conference on 15 February 1955, sending a shock wave through the gemstone industry. De Beers stock dropped dramatically, while General Electric stock climbed. The diamond industry was in turmoil.[2]

General Electric however almost made a similar mistake to that of ASEA. Initially patents were registered only with the US patent office, and a secrecy order imposed, primarily to enable the improvement of the process before GE applied for an international patent. During this time however, De Beers, already well known as the consortium controlling the international natural diamond industry, heard of the news of the General Electric success.

The prospect of synthetic diamonds flooding the lucrative gem diamond market was naturally very unsettling for De Beers, who had established their own synthetic diamond development ventures at the De Beers laboratories located in Johannesburg, South Africa. The newly established research division tried numerous methods in attacking the problem. Independently of the GE team, De Beers designed and developed a high-pressure device very similar to the belt apparatus invented by Hall at GE. By 1957 GE had begun marketing some of their synthetic diamonds. Chemical analysis of these samples gave the De Beers team a chemical advantage.[2]

The De Beers team completed their high-pressure device in May 1959.[2] Following their inevitable success, they quickly began preparing a patent application. Unfortunately, just days before commencement of the De Beers patent proceedings, General Electric filed their international patents. General Electric and De Beers immediately went to court to decide the future of the De Beers synthetic diamond device and project.[2]

The litigation lasted for six years and included testimonies from many distinguished scientists, including the controversial testimony of UCLA's George Kennedy, widely known in the field of high-pressure physics. The court proceedings resulted in De Beers purchasing limited rights to the belt apparatus in the form of a licensing agreement.

Megadiamond Corporation

In 1955, a few months after the invention of the belt apparatus at General Electric, Tracey Hall left the company. He had been unhappy with GE and the Project Superpressure for some time, mainly for personal reasons that were compounded by the inter-team rivalry experienced between himself and the remaining team members.[32]

Upon leaving GE, Hall became a professor of chemistry at the Brigham Young University in Utah. In his new capacity he was immediately struck with the problem of not being able to use his new belt apparatus, the patent for which was held by General Electric. After a number of futile attempts at persuading the Department of Commerce and General Electric to allow him limited use, he decided to invent another device.[2]

There was another obstacle in his way now, not only must the device be capable of the high pressures and temperatures equal to that of the belt device, but it must not conflict with the belt design either. Within two years of research and development he succeeded in this aim. The result was the tetrahedral press, funded to a sum of $10 000 from the Carnegie Institute in Washington (Figure 5.12).[2]

The tetrahedral press adopted the same principles, of the massive support tapered anvils, the non-extruding gasket and electric heating, but was quite different in every

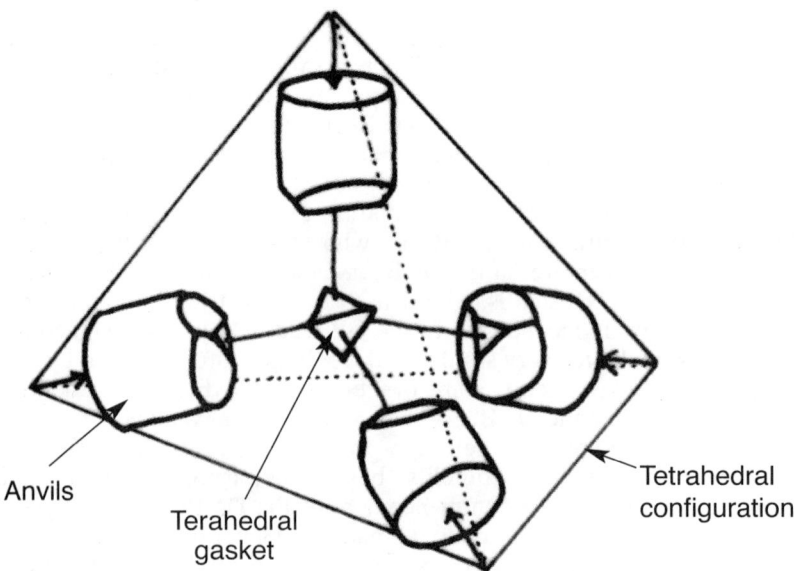

Figure 5.11 Schematic diagram depicting anvil configuration of the tetrahedral press.

Success in diamond synthesis 41

Figure 5.12 Tetrahedral press invented by H. Tracy Hall at Brigham Young University's high-pressure laboratory. (Courtesy H. Tracy Hall Foundation © 1998.)

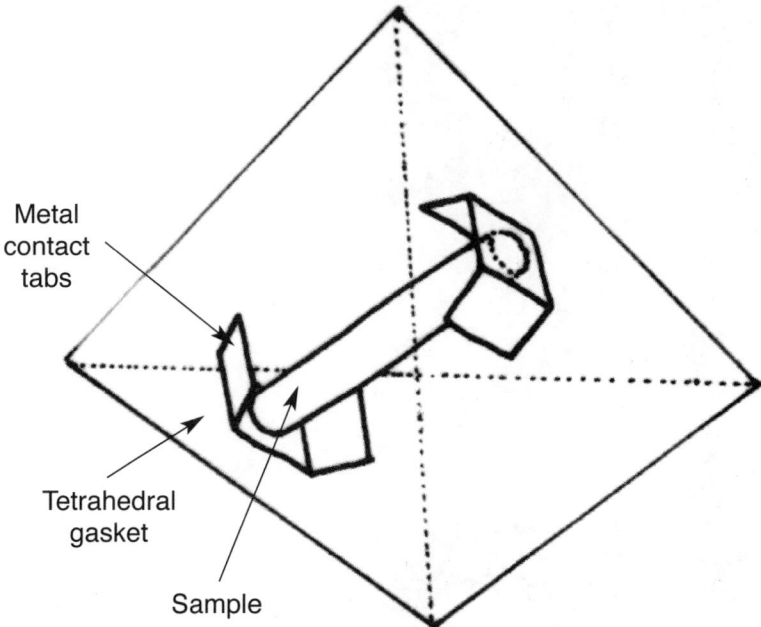

Figure 5.13 Schematic diagram of the reaction cell of the tetrahedral press.

42 The diamond formula

other regard. Rather than following the piston-in-cylinder theme, the tetrahedral press adopts more of the standard tapered anvil stance. Hall did not know at the time that it more closely resembled the ASEA device than any other apparatus.

The name 'tetrahedral press' provides a clue as to the structure. It comprises of four piston shaped anvils, arranged as the arms of a tetrahedron. The principle behind the arrangement of the anvils had most to do with the simplest and most elegant way to apply equal pressure on the sample from all sides. The belt apparatus applied pressure to its sample from only two directions; up and down. The application of pressure from all directions, as performed by the tetrahedral press, most closely resembles the conditions for formation of natural diamonds deep within the earth.[6]

The chamber created by the intersection of the anvils is, by definition, in the shape of a tetrahedron. A gasket for such an arrangement is difficult to achieve; however, Hall devised an ingenious way of imbedding his sample within a pyrophyllite tetrahedron that was 25 per cent larger than the chamber created by the anvils. By so doing the slightly larger pyrophyllite tetrahedron deformed and produced the seal. This allowed the anvils to contact with metal tabs protruding from the ends of the enclosed graphite sample. An electric current passing through the anvils heated the sample via the metal tabs.[6]

Hall deliberated on the design of his tetrahedral press for some time. He had to be absolutely sure that the invention was unique and in no way an infringement on the belt apparatus. Upon deciding that the concept was original he applied for a patent. The Commerce Department applied a secrecy order to the tetrahedral press, but subsequently lifted it by 1959, allowing Hall to finally bask in the glory of his success and inform the world.[2]

Figure 5.14 Close view of one of the first cubic six anvil presses. (Courtesy H. Tracy Hall Foundation © 1998.)

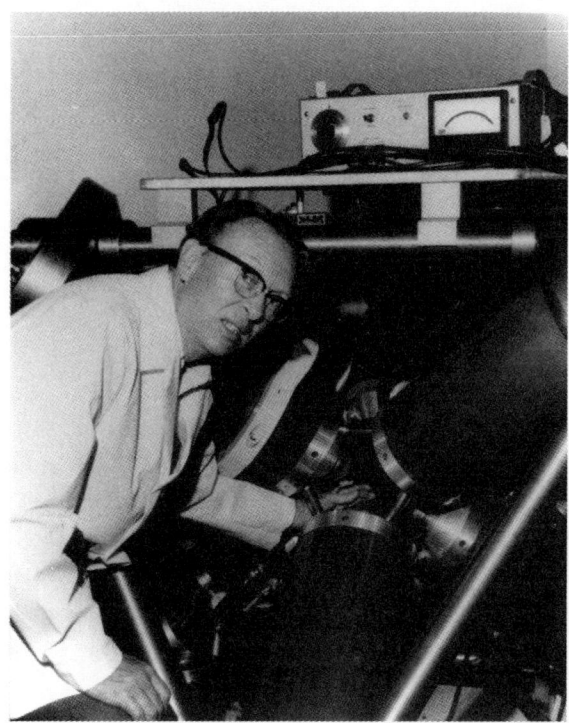

Figure 5.15 H. Tracy Hall placing a sample in the cubic press. (Courtesy H. Tracy Hall Foundation © 1998.)

In 1966 Hall founded the company Megadiamond Corporation, now called Megadiamond Industries. The company continues to produce synthetic diamonds and related products such as synthetic carbonado as solid parts, under the trade name Megadiamond®.[6]

In addition to the belt apparatus and the tetrahedral press, Hall is responsible for the invention of the six-anvil cubic press. This design was conceived in the early 1960s and is currently used by many smaller scale operations such as the US Synthetic Corporation in Provo, Utah.[2] Hall has now retired from active participation in the company Megadiamond. As holder of some 20 patents he still devotes much of his time to research.[6]

Russian diamond makers

History progressed and more diamond manufacturers entered the industry. The Russians were among these, employing a device all of their own. Although this is a modern enterprise in comparison with the previous outlined inventions, it is included here as a description of another high-pressure/high temperature apparatus and method. The Russian synthetic diamond producers, while integrating the current technology into a unique apparatus, have still used the flux method employed by the previously mentioned manufacturers. Regardless of technology, device and innovation the need for high pressure, high temperature and a catalytic metal solvent (or flux) are consistent. Internationally the belt apparatus is by far the most popular device for the economic production of synthetic diamonds, but the Russian researchers have devised

44 The diamond formula

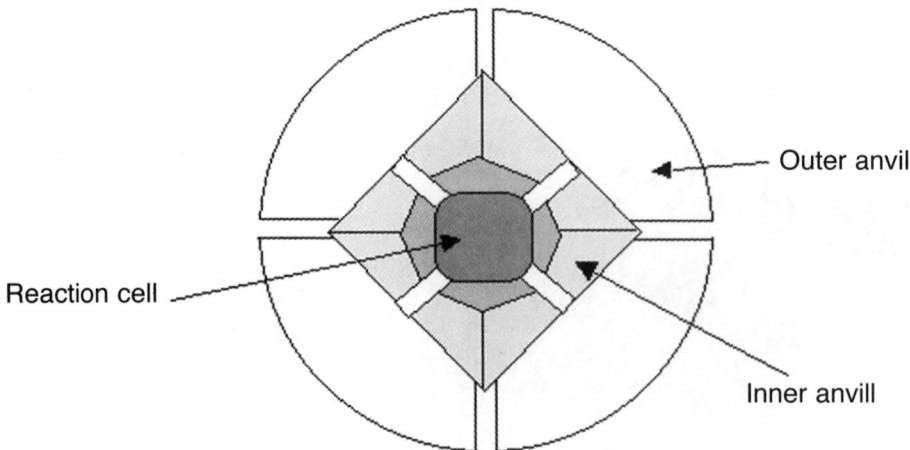

Figure 5.16 Schematic diagram of the interior of the BARS device including the multi-anvil design with internal cavity. (Courtesy *Gems and Gemology* **29**, 4 © GIA 1998.)

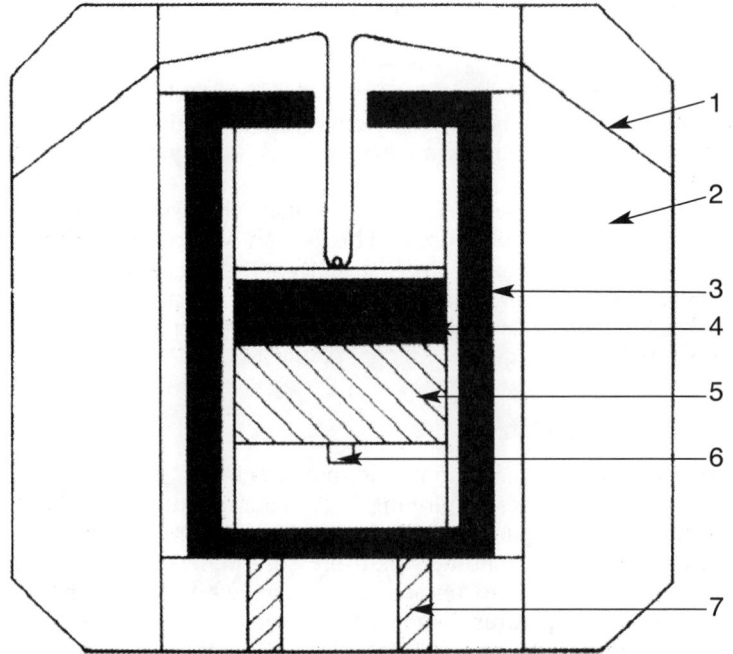

Figure 5.17 Schematic diagram of the reaction cell of the BARS device. Components numbered and outlines in corresponding text. (Courtesy *Gems and Gemology* **29**, 4 © GIA 1998.)

an unusual system, and are having great success. Additionally the Russian researchers have crossed the line from academic development to full scale production.

At Novosibirsk in Russia, scientists utilize a device that adopts the principals of the split sphere system of ASEA. The device is termed the BARS apparatus.[33] The term 'BARS' is an acronym for Bespressovye Apparaty tipa Razreznaya Sfera (or translated to English, 'split sphere no press apparatus').[34] It consists of a series of anvils arranged in a spherical shape, in the same way as the ASEA split sphere system,[28] with decreasing number of anvils in concentric layers.[33] The outer layer of anvils comprises eight, with the external appearance of a sphere and the internal shape of an octahedron. The second set of six tungsten carbide anvils fit neatly within this cavity, forming a still smaller cube-shaped cavity within the octahedron. Here, at the centre of the high-pressure device, the reaction cell is housed. The complete arrangement is placed within a press, and the pressure evenly distributed among each of the anvils. This effectively results in even and uniform pressure being applied to the entire reaction cell.[33]

The cell itself is a cubic shape and is smaller than the more traditional reaction cell of the belt apparatus. It incorporates a thermocouple on the outside, an electric heating component, and utilizes a seed crystal to instigate synthesis. Applied to the cell are pressures around 55 000 to 65 000 atmospheres, at temperatures in the vicinity of 1350°C to 1700°C, well within the desired environment for successful diamond synthesis.

The cube-shaped reaction cell is a complex arrangement of unique design. The numbered pointers in Figure 5.17 represent the following:

- thermocouple (1)
- pressure transferring medium container (2)
- heating unit (3)
- carbonaceous source material (usually diamond powder) (4)
- catalyst (4)
- seed crystal (5)
- electrical power supply (6).

The catalysts used by the Russian producers with their BARS apparatus include the transition metals iron, nickel and manganese, and their alloys. Total growth is estimated at more than 5 mg per hour.[33]

Following on from the success of these major players in the beginning of the synthetic diamond industry, many other companies, universities and governing bodies have entered the field. The technology once held so secret by General Electric has been modified, adapted and built on to form similar devices to make diamonds all over the world.

It was inevitable once success was achieved that the focus would shift from pure research to pure economy. Synthetic diamonds soon became big business, as we shall see.

Chapter 6

Birth of a billion-dollar industry

Following the patenting of the diamond synthesis procedure and apparatus, GE went into production of their new synthetic industrial-quality diamond grit almost immediately, in 1957.[2] Initially applications for five patents were submitted in 1955. These are outlined briefly in Table 6.1 with a short description of the processes and materials involved.

Marketing of the synthetic diamond grit began in 1958, after the patent applications were submitted during the years 1955 to 1957, but before the patents were granted in 1960.[2] It is from this release that De Beers obtained sample synthetic diamonds and gained vital chemical information with regard to the product and process. This effectively saved them years of experimentation that would have been otherwise necessary.[2] Since this time General Electric have significantly increased the number of specialized products, and other companies have entered the synthetic diamond industry.

Since the 1960s synthetic diamond production has increased rapidly. In 1980 the estimated annual world production was approximately 100 million carats.[6] Recent production levels of synthetic diamond grit has been estimated at around 500 million

Table 6.1 First patents for the synthesis of diamond granted to General Electric in 1960.

Date granted	US patent number	Patent subject
21 June 1960	2,941,248	For apparatus to define a reaction of controllable dimension, localized pressures and temperatures that may be maintained for desired periods. **(Belt apparatus)**
2 August 1960	2,947,608	Improved diamond synthesis method eliminating need for determining temperature at point of conversion and offering indication as to the moment of conversion. **(Resistance inflection)**
2 August 1960	2,947,609	Conversion achieved at low pressure of 50 000 atmospheres through addition of catalysts (alloy of Cr, Ti and Mn) at temperatures of 2000°C to 1200°C.
2 August 1960	2,947,610	Non-diamond carbon to diamond carbon conversion using one of eleven metal catalysts.
2 August 1960	2,947,611	Method using platinum as a catalyst at 2050°C to 2500°C to enable conversion of a variety of carbonaceous materials.

carats per annum, representing approximately 80 per cent of the industrial diamond grit industry.[1]

Countless companies have entered the market place and production facilities are spread across the globe. It may never be fully appreciated just how many operations have incorporated diamond synthesis into their portfolio, due to the access to information in some countries. Hall himself has recalled during a tour of China in 1988 visiting a small family-owned jade carving operation and finding a 1000 ton press (in a backyard shed), used to synthesize diamonds for carving tools.[2] The estimated world market for synthetic diamonds is approximately one billion (one thousand million) dollars.

Using the United States of America as an economic model for the consumption of synthetic diamond versus natural diamond, we can see that synthetic diamond production has maintained a record high level over the course of many years. Most industrial diamond produced domestically within the USA was, during this time, synthetic grit and powder. Seven firms recovered and sold industrial diamond, both natural and synthetic, as the principal product. About thirty-five firms recovered industrial diamond in secondary operations.

Major uses of all industrial diamond were:

- machinery, 27 per cent
- mineral services, 18 per cent
- stone and ceramic products, 17 per cent
- abrasives, 16 per cent
- contract construction, 13 per cent
- transportation equipment, 6 per cent
- other, 3 per cent.

The mineral services industry and primary drilling accounted for 59 per cent of diamond consumption during the time of the survey. About 26.4 million carats of natural industrial diamonds were salvaged in secondary production from salvage stone, sludge, and swarf.

Sources of diamonds imported into the USA during the period 1991 to 1994 including bort, grit, powder and dust, both natural and synthetic, are given below.

Synthetic industrial diamond imports by countries of origin:

- Ireland, 63 per cent
- China, 7 per cent
- Russia, 7 per cent
- other, 23 per cent.

Natural industrial diamond imports by countries of origin:

- United Kingdom, 30 per cent
- Zaire, 23 per cent
- Ireland, 16 per cent
- other, 31 per cent (including Australia).

One point that becomes evident from these statistics is that the huge US market consumes more synthetic diamonds than they can produce. This information was compiled by the Industrial Diamond Specialist, United States Department of Commerce,

Table 6.2 US consumption statistics for industrial diamond. (Source: Industrial Diamond Specialist, United States Department of Commerce, Mineral Commodities Summaries January 1996.)

Consumption statistics	1991 USA	1992 USA	1993 USA	1994 USA	1995 USA
Bort, grit, powder and dust (synthetic diamond, millions of carats					
Production, synthetic	90.0	95.0	105	104	115
Production, secondary	3.5	3.4	15.9	16.0	26.1
Imports for consumption	70.0	97.3	133	174	200
Exports and re-exports	78.8	83.6	107	153	120
Sales from Gov. stockpile	5.0	10.4	–	2.0	0.2
Apparent consumption	89.1	122	146	141	222
Value of imports ($ per carat)	$0.83	$0.70	$0.61	$0.51	$0.46
Import % of consumption	Est.	19%	18%	15%	36%

Mineral Commodities Summaries January 1996. What these figures fail to capture is the diversification of the industries involved in manufacturing and consumption.

Diversification has been an international trend in the synthetic diamond industry. Organizations delving into the synthetic diamond field have hailed from countries such as Japan (notably Sumitomo whose contribution to diamond synthesis will be discussed later), South Korea, Russia (also to be discussed later), China, the French company Du Pont and numerous private companies throughout the USA.

Additionally synthetic diamonds have become big business for government bodies,

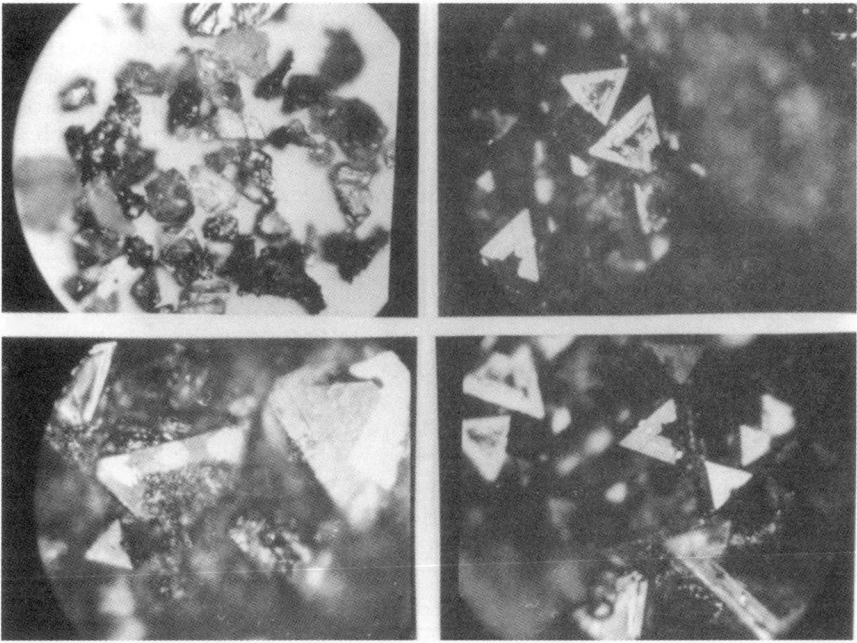

Figure 6.1 Different samples of the first synthetic diamonds under magnification. (Courtesy H. Tracy Hall Foundation © 1998.)

with departments such as the United States Department of Defense funding synthetic diamond projects for application in the Star Wars Initiative.[35]

Competition for funding allocated for research into diamond synthesis and the synthesis and development of other super-hard materials is heated, with the private and public sectors blurring, overlapping and collaborating in a way not encountered in other fields.

Probably among the most important developments since the initial success have been the development of the processes for production of synthetic diamond films and coatings (discussed in Chapter 14), and synthetic diamond mono-crystals. The latter was achieved, again by General Electric, announced on the 28 May 1970,[5] and published in 1971.[6] At this time General Electric reported being able to produce gem-quality crystals up to two carats in size.[2] To achieve this, modifications were made to the reaction cell and seeds were introduced. The new arrangement encouraged the diamond to be deposited onto the seed crystals via a thermal gradient (see Chapter 7). This represents only a slight overall change to the existing method of synthesis.

Since the advent of successful synthesis of large gem-quality diamond crystals, many companies have emerged as active participants in this area. Sumitomo Electric Industries of Japan was the first to offer a commercially viable product of this type, marketed as Sumicrystal® in 1985. Sumitomo's research facilities are housed at Itami, Japan, where the company specializes in the commercial production of synthetic gem-quality mono-crystals. These are available in a range of sizes and prepared forms, including a product category guaranteeing synthetic diamonds of one carat. Sumicrystals® are also available up to two carats in size, though again all products are sold pre-prepared.[36]

Sumitomo claim their product is highly thermally conductive, fracture resistant and of high purity. These commercial products are yellow in colour, although recently Sumitomo have claimed to be able to produce near-colourless crystals. The price of the product varies from US$60–$140, for prepared (laser cut and partly polished) pieces ranging from 0.10 ct to 0.40 ct.[37]

Sumitomo have a stated policy that all of the company's synthetic diamond products are sold as tabular, finished pieces, and are only for industrial purposes such as heat sinks and electronic components.[38] Despite this, we will discuss the properties of these synthetic diamonds in the next chapter as this situation could change in the future.

De Beers and GE also produce large gem-quality crystals, the majority designated for commercial applications, and the remainder for experimental purposes. The De Beers operation includes enormous facilities in Rand, South Africa,[6] a new facility in the Isle of Man[2] and at Shannon, Ireland, opened in 1963 in conjunction with the Zairian mining company Société Minière du Beceka, operated by Ultra High-pressure Limited.[5]

These facilities account for approximately 50 per cent of the world's synthetic diamond production. In addition to this De Beers held a 50 per cent interest in the operations at ASEA in 1965,[39] and took complete control of the Scandiamant division in 1975.[2] De Beers were among the first to design and introduce computerized microprocessors to accurately control the temperature and pressure of the (belt type) devices. These devices expose the samples to 1500°C and 60 000 atmospheres for only a few minutes to obtain successful results.

Prior to this, a number of smaller scale operations have achieved diamond synthesis by alternative methods. For example the Swiss biochemist D.L. Tomarkin together with the Puerto Rican chemist M. Vilella have patented a furnace to produce small diamonds at high pressures of a reported 3 675 000 atmospheres, and high temperatures of 3300°C.[5] In Holland the N.V. Bronswerk laboratory had produced diamonds as early as 1955, in a method closely related to that of GE. In 1961 the National Physical Laboratory in the United Kingdom achieved limited success.[5]

Evidence exists that synthetic gem-quality diamonds were produced in Russia as early as 1967. During this year Professor Bakul, Director of the Kiev Synthetic Diamond Research Insititute, representing a Russian delegation of unknown origin or nature, approached a Belgian diamond cutter named M. Bonroy. He was asked to cut some synthetic diamond crystals. The crystals were reported as light yellow in colour, of an extraordinary shape and of high purity.

The crystals also proved to be harder than natural diamonds and could not be sawn in the usual manner. Consequently Bonroy was invited to speak on his findings during the cutting process at a Kiev symposium in 1971. This was the first public announcement of the successful synthesis of diamonds by a Russian group.[5]

There have since been other reports on Russian synthetic diamonds entering the jewellery industry as faceted gemstones, originating both from known and unknown sources. Properties of these synthetic diamonds are covered in the next chapter. In 1980 the Russian production level of synthetic industrial diamond was in the vicinity of 15 million carats annually.[39]

Tairus, the Russian synthetic gem manufacturer, built on the synthesis of other gems such as ruby and emerald, now offer synthetic gem-quality diamond for sale and distribution, as crystals and faceted specimens world wide.

Colourless gem-quality synthetic diamonds are also marketed and distributed by the American company, Chatham, also famed for the manufacture of the widely used synthetic rubies and emeralds. Synthetic diamonds were first offered to the public by Chatham at the Las Vegas jewellery show in 1993, apparently from a Russian source,[37] and not from Sumitomo, GE or De Beers.

In addition to the increase in the number of companies pursuing a fortune in the diamond synthesis field, manufacturers of the high-pressure devices have also increased. Large and cumbersome machinery is used by many institutions.

Spare parts are also a very important part of the high-pressure industry. The strain placed on components of these enormous contrivances causes many to fail, and replacements are essential. The cost of these replacement parts (such as the replacement of the gasket in Figure 6.2) constitutes a major economic burden. It would be fair to say that a considerable percentage of production costs for large manufacturers can be attributed in some way to the failure and replacement of machine parts.

Applications and products

Natural diamonds have held a special place in industry for decades. Their innate hardness, thermo-conductive properties and resistance to chemical attack have made them useful in many areas. They have become primary tools for a whole range of cottage industries that have grown to become million-dollar enterprises in their own right. Industrial diamonds are integral in the machining, grinding, cutting, sawing and polishing of other super-hard materials. Additionally, diamonds' hardness lends them to drilling and to drill bits, from those used on oil rigs to the type available at the local hardware store. Diamonds are also used widely as heat sinks in electronic equipment. Their un-wettable nature proves perfect for ultra sharp surgical instruments. These are samples of the countless uses for industrial diamonds. It is easy to see why the race to synthesize diamonds was so heated.

52   The diamond formula

Figure 6.2 The Belt device after a run, showing the deformation of the gasket. (Courtesy H. Tracy Hall Foundation © 1998).

Synthetic diamonds have been integrated into the existing market, and due to their superiority in many areas, have opened up new fields. Their superiority has its base in the control of their physical properties during synthesis. For example, in order to achieve a uniform grade and size of diamond grit for the purpose of grinding a carbide machine part, natural diamonds must be crushed.

This crushing process, while ensuring a uniform size of grit, leaves behind cleavage traces in each tiny diamond piece. These cleavage traces represent an effectual defect in the grit, meaning that the diamonds will be less effective in the grinding process because the sharp corners needed to score the carbide are lost as the process continues.

Synthetic diamonds can be produced in any predetermined shape and size, which means that the usual crushing is not necessary. Without crushing the industrial synthetic diamond grit has no planes of weakness and retains the sharp edges and corners for which it is prized. It was found that the grit can be engineered by monitoring and modifying a number of parameters. These are mainly the length of time of the synthesis procedure, the exact heat, the exact pressure, the catalyst used, and the nature of the starting carbonaceous material. As we will see in Chapter 7, the higher the temperature and pressure conditions during synthesis the more the morphology of the resulting synthetic diamonds will lean toward an octahedral habit. Naturally the longer the run the larger the synthetic diamonds, as the carbon has more time to crystallize. This also means that these crystallites have time to grow in size at the expense of further nucleation.

Birth of a billion-dollar industry   53

The discussion in Chapter 7 relates mainly to the production of synthetic diamond mono-crystals. The principles are the same as those applicable when monitoring the size and shape of synthetic diamond grit. More information on a new area of synthetic industrial diamond production, in the form of coatings, films and laminates is also provided in Chapters 14 to 16. This field has opened doors to new applications for the hardness of diamond, and its optical, chemical and thermo properties. This technology gives a whole new meaning to the concept of being able to control the form of synthetic diamonds.

Today GE offer to industrial diamond consumers a wide range of synthetic diamond products designed and synthesized especially for different purposes. Table 6.3 is a list of some of the products marketed by General Electric and the associated trading names.

This product list has been divided into categories according to their market niche. It includes a brief description of their intended purpose (courtesy of General Electric). The larger diamonds are used for sawing concrete, granite, and marble. Smaller diamonds are used in grinding wheels.[40] It should be noted that none of these purposes include faceted synthetic diamonds designed for the gem industry. In addition to the existing products, General Electric conduct on-going research into the development of new applications for synthetic diamonds, the appropriate gauge of product required, and improvements possible for standard lines.

Table 6.3 Examples of General Electric synthetic industrial diamond products. (Source: General Electric official web site.)

GE category	GE product	Designed for:
Grinding and polishing	Man-Made® Diamond	Manufactured diamond crystals designed for grinding nonferrous, nonmetallic, and ceramic materials.
	Micron® Diamond	For polishing and super finishing of ferrous and nonferrous materials.
Machining	Compax® Diamond	Polycrystalline diamond blanks in a wide range of shapes and sizes that are fabricated into tools for machining nonferrous and nonmatellic materials.
	BZN® Compacts	Polycrystalline blanks that are fabricated into cutting tools for machining steel, cast iron, superalloys, and other ferrous materials.
Wire drawing	Compax® Diamond Die Blanks	Self-supporting and tungsten carbide supported polycrystalline diamond for wire drawing applications.
Sawing and drilling	MBS® Diamond	Diamond saw blades, wire saws, and core drills are used to quarry and shape natural stone and mastonry and to renovate concrete highways, bridges, buildings dams and power plants.
Oil and gas exploration	Stratapax® drill blanks	Used to improve production in the drilling industry. The self-sharpening cutting edges of the blanks allow for more rapid penetration than with traditional crushing or fracturing methods.
	Geoset® drill diamond	In conjunction with Stratapax.

54 The diamond formula

A typical production cycle will yield approximately 300 carats of synthetic industrial diamond of various grades.

General Electric are not, however, the only company with an extensive range of synthetic diamond products. De Beers produce around 20 different synthetic diamond products in addition to their range of natural industrial diamond, for which they have virtual control. De Beers began marketing their synthetic diamond mono-crystal products in 1987.[41]

De Beers pride themselves on the assurance given to all their synthetic diamond clients that each run of all products is fully tested for mechanical properties in their Grit Processing Department. Here the grit is inspected and exposed to the rigors of in-use wear to ensure stability, suitability and structural integrity. In this way they can be sure of maintaining a high level of quality control. Table 6.4 shows some of the major product lines offered by De Beers to the industrial diamond buying public.[42]

There are many other companies offering synthetic industrial diamond products, either in the capacity of manufacturer, distributor or both. The products from GE and De Beers tabled here provide a sampling of the types of synthetic diamond grit products available in the open market place. Synthetic diamond grit ranges in price from about ten cents per carat to four dollars per carat, the price depending on size, shape, and quality.

Table 6.4 Examples of De Beers synthetic industrial diamond products. (Source: *Indiaqua*.)

De Beers category	De Beers product	Features and applications
Synthetic metal bond (SDA series)	SDA100S	Extra high strength cubo-octahedral grit. Exceptionally clear and free from impurities.
	SDA100	Extremely high strength cubo-octahedral grit. Designed for sawing and drilling stone and concrete.
	SDA100+	Smaller grained version of SDA, with a higher technical specification.
	SDA85	Intermediate strength abrasive for sawing of medium density stone materials.
	SDA85+	Intermediate strength abrasive grit with a higher technical specification.
	SDA	Strong, blocky synthetic diamond grit, designed for free-cutting and long lasting.
	SDA+	Strong, blocky grit with a higher technical specification.
	SDAT	High strength titanized (coated) grit for concrete sawing.
Synthetic metal bond	EDC	Engineered synthetic diamond grit, irregular in shape, with excellent bonding and abrasive qualities for saws and tools.
	MDA100	Ultra-strong shaped grit with enhanced impact strength for grinding (wet) ceramics, glass, and refractory materials.
	MDAS	Very strong cubo-octahedral grit with extremely high bond retention for grinding (wet) ceramics, glass and stone.
	MDA	Friable alternative to MDAS for grinding (wet) carbides glass and ceramics.

Table 6.4 (*continued*)

De Beers category	De Beers product	Features and applications
Synthetic resin bond	DXDAMC	Metal (55% total weight nickel) clad synthetic diamonds grit for grinding (wet) metal and carbides.
	CDA	Mosaic structured, friable, blocky grit for grinding tungsten carbide.
	CDA55N/30N	Metal (55% and 30% total weight nickel respectively) clad synthetic diamonds grit for grinding carbides.
	CDA50C	Metal (50% total weight copper) clad synthetic diamonds grit for grinding (dry) carbides.
	CDA-M	Metal (multi-grain) clad synthetic diamonds grit for grinding (dry) carbides, and finishing operations.
Synthetic roller dresser	SRD	Well formed cubo-octahedral synthetic diamond grit processed to give shape edges for toolmakers and rotary dresser products.
Synthetic micron	Micron MDA	Synthetic diamond powder for fine grinding, lapping and polishing.
	Micron CDA	Mosaic structure, friable strong powder for fine grinding, lapping and polishing.
Syndrill		Various sizes and shapes. Solid drill bits of synthetic diamond and tungsten carbide, designed for rock cutting.
Synda X3		Synthetic diamond polycrystalline thermally stable inserts. Available in rectangle (L series), triangle (T series), pentahedron (P series) and hexahedron (H series).
Syndie		Polycrystalline synthetic diamond wire drawing die blanks, available supported or self-supported in various sizes.
Syndite	R series (PCD)	Round polycrystalline synthetic diamond tool blanks in various sizes.
	T series (PCD)	Triangular polycrystalline synthetic diamond tool blanks in various sizes.
	L series (PCD)	Rectangular polycrystalline synthetic diamond tool blanks in various sizes.
	PLUS (PCD)	Round polycrystalline synthetic diamond tool blanks in special sizes.
	Microdrill blanks	Specialized small drill polycrystalline synthetic diamond drill blanks.
	Macrodrill inserts	Two styles (Pentahedron and Chevron) polycrystalline synthetic diamond drill inserts.

Part 2

Synthetic Diamond – Friend or Foe?

Chapter 7

Synthetic diamonds

In this chapter we will discuss the modifications and processes designed to achieve synthetic diamond gem-quality mono-crystals, then move on to explain how the alteration of these factors can determine the desired properties of the resulting crystals. This includes the aspects of colour in synthetic diamonds and the manipulation of colour. The remainder of this chapter will be devoted to describing synthetic diamonds, their morphology and unique properties.

Growing gem-quality synthetic diamond mono-crystals is a very difficult process. When using the term mono-crystal, reference is being made to larger synthetic diamond crystals usually of the quality one would expect from a naturally occurring gem-quality diamond. Their size and clarity make them suitable for faceting, and the growth process allows for single crystals to be grown independent of other crystals. This is one of the many aspects of the technique that differs greatly from the production of synthetic diamond grit, outlined in Chapter 6. The first and most obvious difference is the length of the actual procedure.

Synthetic diamond mono-crystals require longer growth periods to evenly deposit the diamond-layer over the growing crystal. Consider the growth of a pearl, with the concentric layers of nacre evenly deposited around the seed. Eventually a pearl forms, and as the nacre continues to be deposited the pearl grows larger and more beautiful. This analogy may be applied to the growth of synthetic diamond mono-crystals. As the synthetic diamond grows the diamond carbon is deposited in layers. Unlike pearls however, the layers are not concentric (see later sections) and the growth is rather quick in comparison to a pearl, that may take years to form. Growth rates of synthetic diamonds vary from approximately one 5 mm crystal per week,[43] to half a carat per day, per crystal.[44]

Another area of the synthesis procedure, where this method differs from that used to make synthetic diamond grit, is in the uniformity of the process. Conditions for the growth of mono-crystals must be kept extremely uniform and are maintained at the required levels for extended periods. Both the applied pressure of the device, and the temperature required to enable phase transition must not fluctuate. Any fluctuations in the temperatures or pressures could result in spontaneous nucleation of the carbon material, at the expense of the diamond deposition on the seed crystal.[6]

As depicted in Figure 7.1, seed crystals are introduced into the process. The seed may be a slice of either a natural diamond or a previously grown synthetic diamond, and the feed or source material (usually synthetic diamond powder) is encouraged to grow as layers onto the seed. This is achieved by the positioning of the seed in the reaction cell. Referring again to Figure 7.1, the seed is placed at ends of the reaction

60   The diamond formula

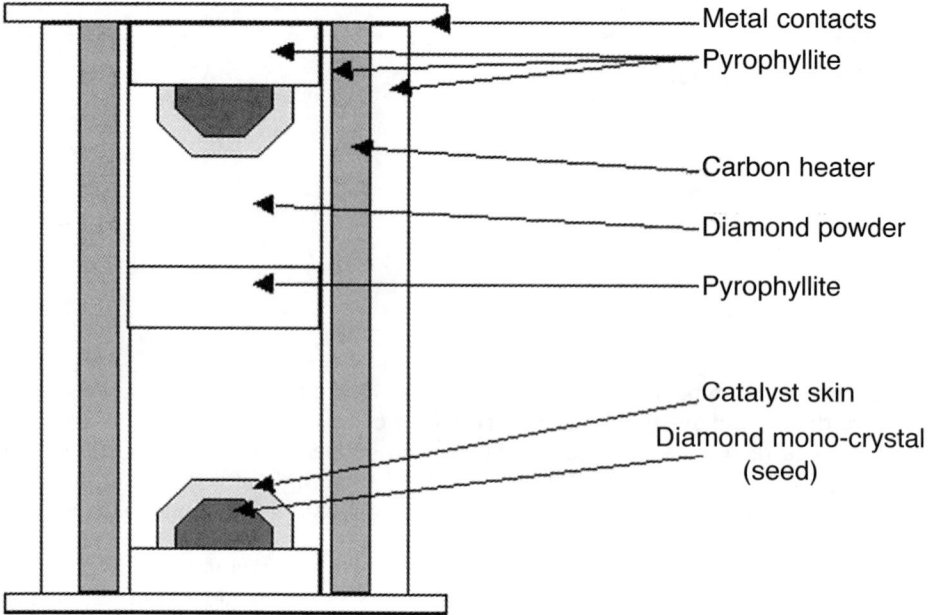

Figure 7.1 Schematic diagram of the reaction cell (Belt device design) modified for the production of synthetic diamond mono-crystals.

cell with the carbon source opposite, as this produces the thermal gradient that is optimum for deposition.

Although graphite is suitable as the starting material for synthetic diamond mono-crystals, it is not the material of choice. The level of pressure and temperature required to ensure phase transition in graphite are high. For the purpose of producing mono-crystals, fine diamond powder is used in the place of graphite. This is an added precaution against problems in non-diamond-carbon to diamond-carbon conversion. Such problems are caused by volume shrinkage as the graphite collapses to the more dense structure of diamond, which results is a loss of pressure in the reaction cell.

This can result in graphite inclusions subsequently reducing the purity of the synthetic crystals. Many commercial applications require a high level of purity within the crystals, and such inclusions would render them useless for their prescribed high-tech purposes. Additionally, the working volume of the reaction cell inside the belt apparatus for synthetic diamond mono-crystals has been increased to approximately 10 cm in diameter and about 15 cm in height.[44] This offered advantages in ensuring uniform growth and commercial viability of production.

To further increase the commercial viability of each run, the reaction cells are divided into layers, each acting as an individual reaction cell in its own right. The advantage of this is clear, as each 'cell' or layer can house a separate crystal. Hence one device can produce a larger number of crystals per experiment in the time allowed. The layers are divided by pyrophyllite to maintain the integrity of the unsupported seal.

A thermal gradient of around 30°C is induced[1] to dissolve the carbon. The catalyst around the seed is super-saturated with carbon and therefore deposits on the seed (which has a lower temperature), while the catalyst in the region opposite the seed is under-saturated and continues to dissolve more carbon. This in turn feeds the deposition, and so on. The metal contacts at the ends of the reaction cell are subject to heat loss, causing a lower temperature at the ends of the cell, and conversely a higher temperature toward the centre, where heat loss is much less.

This apparently causes the molten metal catalyst around the growing synthetic diamond mono-crystal to move inward, toward the hotter part of the cell. The result of this mobile catalyst is that the dissolved carbon is deposited in its wake, and crystallizes as diamond on the seed. If spontaneous nucleation occurs this steady deposition will be interrupted and additional crystals will form in the catalyst around the seeded mono-crystal, in the same way that grit is formed.

New advances are being achieved in the modification and improvement of high-pressure apparatus all the time. The main purpose of such research is to improve the quality and/or the commercial viability of the synthetic diamond crystals produced. More recently researchers and industrialists have made changes not only to the devices but also to the overall technique and methodology of making diamonds.

For example, a new two-stage method has been developed together with the associated technology, designed to dramatically improve the uniformity and quality of synthetic diamond mono-crystals with a higher net growth rate. This new approach adopts changes to the high-pressure reaction cell which includes a recess at the lower temperature side housing the seed crystal at the lowest point. When the recess is filled by the first-stage growth of diamonds from the seed, its upper surface serves as a large seed for the second-stage growth. If the size of the recess is selected appropriately, the perfection and quality of the second-stage is very high. The growth rate is small for the first-stage growth in the recess, but becomes larger for the second stage growth.[45]

The advantages of this new technology are readily evident.

Classification of synthetic diamonds

The majority of synthetic diamond industrial-quality grit and synthetic diamond mono-crystals produced are type Ib diamonds. Synthetic diamond mono-crystals can be achieved in the other types, IIa and IIb (as can synthetic diamond grit if required), as we will discuss later. It is important to note that type Ia synthetic diamonds (usually colourless in natural crystals) cannot at this stage be consistently generated, and any manifestation of this type is an anomaly. The Gemological Institute of America (GIA) research and testing laboratory has reported to date that they have not seen, nor heard of the existence of any synthetic near-colourless pure type Ia diamonds.[46] Without any form of interference the synthetic diamonds will 'automatically' be type Ib.

Type Ib are classified as having nitrogen as an impurity, scattered throughout the lattice as single atoms. The ratio is approximately 1:100 000 carbon atoms.[6] Sumitomo synthetic diamonds for example, contain 30–60 ppm (parts per million) of nitrogen.[37] Nitrogen in the form of a single atom can easily substitute for a carbon atom, as they are both non-metals and chemically very similar. In fact, nitrogen is adjacent to carbon on the periodic table. Nitrogen has an atomic number of 7 (carbon is 6), and an electron configuration of 2-5. Hence a nitrogen atom 'slots' in rather well, with the exception of the extra electron which is left as a kind of loose end.

At the top of Figure 7.2 is a schematic diagram of the diamond lattice (adapted from Bursill and Glaisher, 1985). Note that the white circles represent the carbon atoms and the black circles represent the nitrogen atoms in the lattice. In the lattice to the left (the synthetic type Ib representation), the black nitrogen atoms are spaced

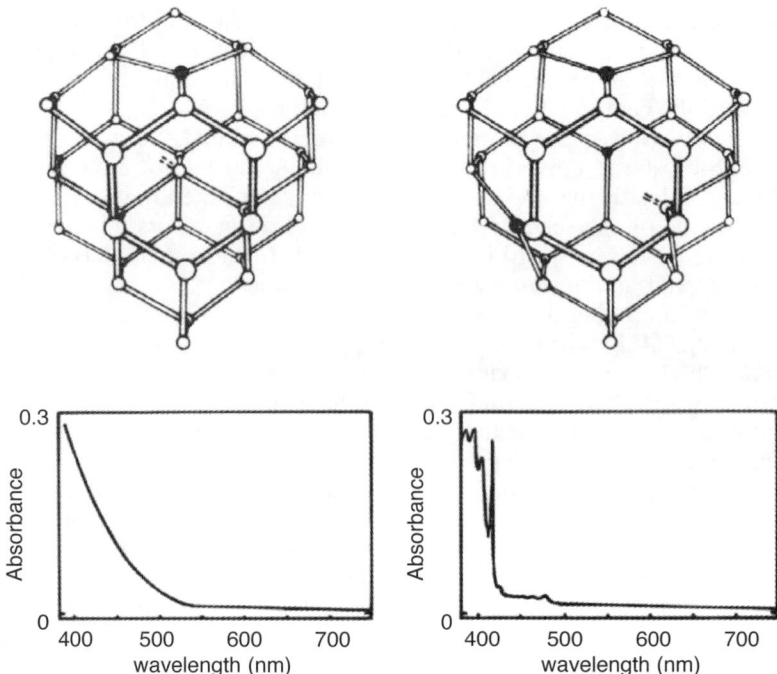

Figure 7.2 Comparison of yellow type Ib synthetic diamond (left) and natural type Ia diamond (right). Features drawing of crystal structures (top) with carbon atoms represented by white circles and nitrogen as black circles. Also shown is the absorption spectra of each type obtained using an ultraviolet-visible spectrophotometer. (Courtesy *Gems and Gemology* **29**, 4 © GIA 1998.)

evenly throughout, replacing what should have been a carbon atom (if the lattice were pure carbon), while in the diagram to the right (the natural type Ia representation) there are marked differences in the atom distribution in the lattice. Notice that at a first glance the lattice appears distorted even before the placement of the black nitrogen circles can be determined. Here the nitrogen forms in clusters (like an N3 centre). The lower half of Figure 7.2 shows the effect this nitrogen distribution has upon the spectra of the respective diamonds. We will discuss spectra in some detail during the course of the proceeding chapters. It is recommended that readers refer back to this diagram at that time.

The cause of colour in synthetic diamond is encapsulated by the wide band-gap theory. Very briefly, lattices with covalent bonding (sharing electrons) can be thought of as producing a band of energy. There are so many atoms packed together that the electrons are collectively 'owned' by the entire lattice, rather than by individual atoms, or groups of atoms.[6]

Consider the structure of an atomic lattice as having band areas, like steps on a staircase (see Figure 7.3a). The lower step represents the low-energy valence band, comprising the electrically neutral lattice, with filled orbitals around each of the atoms. The highest step on the hypothetical staircase represents the high-energy conduction band, which comprises an 'empty' band equating to the high energy levels of

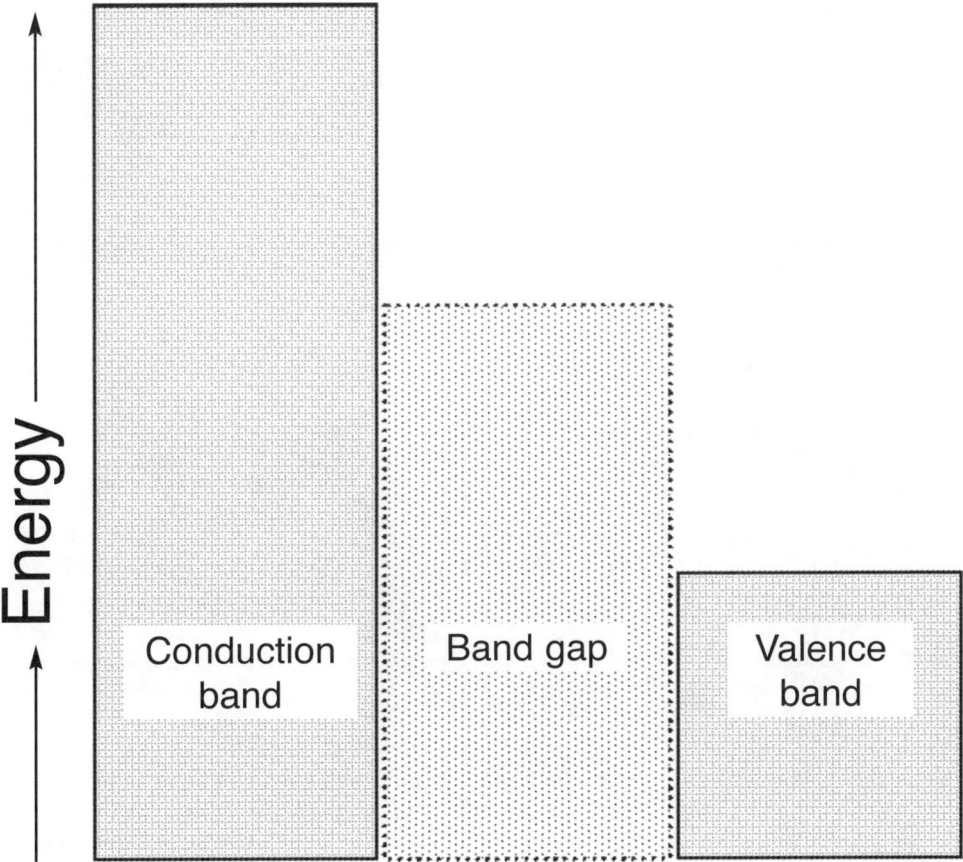

Figure 7.3a Schematic diagram representing energy levels in the valence and conduction bands, and the 'band gap'.

excited electrons. The middle step is also empty, and represents the 'band-gap'. According to the band-gap theory, the extra electron that results from the substitution of a nitrogen atom (called a donor as it has one extra electron to donate) will not fit into the neutral lattice of the valence band. Nor is the extra electron excited by any form of radiation or energy, meaning it has no place in the conduction band. It is merely an 'extra'. Therefore it is found in the 'band-gap'.

The band gap is an unstable position, but as there is no room in the lower valence band, the electron is inclined to absorb energy (from an external source) to excite it into the higher conduction band, where there is plenty of space. For the nitrogen electron in diamond, the energy level required to make this 'jump' equates to frequencies of the visible light spectrum. Higher frequencies of blue and violet are readily absorbed resulting in the transmission of the complementary yellow colour (see Figure 7.3b).[6] This is a very simple and somewhat picturesque interpretation; however, for our purposes it will suffice. For readers interested in further reading in this area there are many fine physics texts defining the topic in more scientific terms.

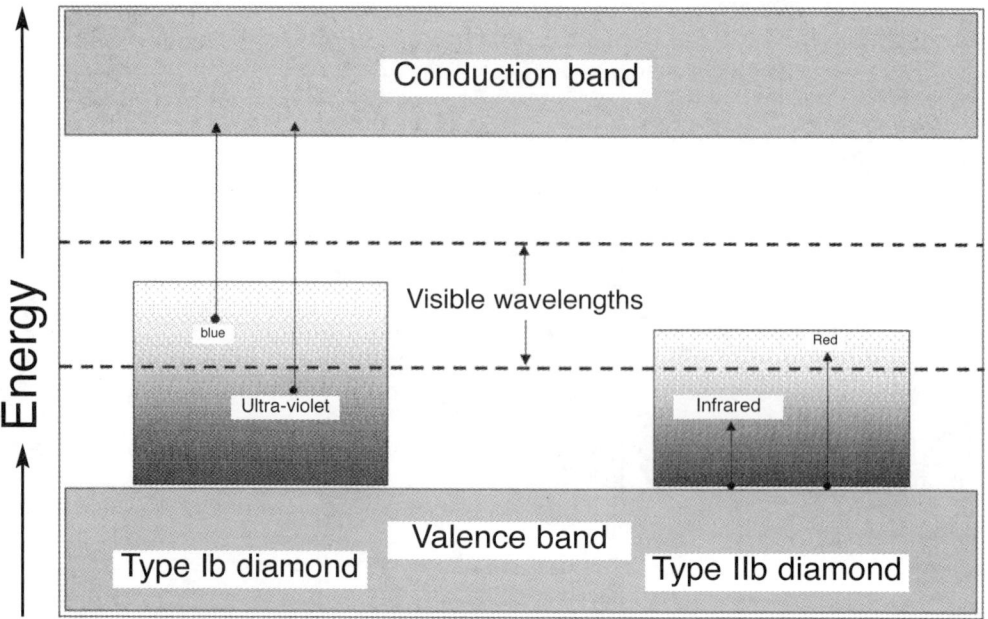

Figure 7.3b Schematic diagram of band theory in pure diamond, showing visible range absorption and transmission for type Ib and type IIb diamond.

This is the reason that all type Ib synthetic diamond will be a rich yellow in colour. The colour cannot be eliminated without removing the extra electrons, which to date has not proved possible. Attempts have been made to treat synthetic yellow type Ib diamonds with the intention of inducing nitrogen clustering (and producing colourless type Ia crystals). However the success of these procedures has not been of the desired magnitude.[33] These treatments and the precedence of nitrogen clustering in synthetic diamonds is covered later in the text.

As mentioned at the start of this section, synthetic diamonds can be produced as type II diamonds. Type II diamonds are very rare in nature (less than 0.01 per cent of natural diamonds occur as type II).[36] They possess enhanced thermal and electrical properties, due to their high level of purity. Type IIb diamonds are the only diamonds that possess the property of electrical conduction, along with their famous and attractive blue coloration. The famous Hope diamond, now on display in the Smithsonian Institute in Washington, USA, is a type IIb natural diamond. The cause of colour, and the electrical conductivity of type IIb diamonds, either natural or synthetic, is also due to the presence of impurities and the band-gap theory.

In type IIb diamonds, no nitrogen is present. Instead boron is substituted into the lattice as an impurity. The ratio is less than 1:1 000 000 ppm carbon atoms.[24] Boron is another element that, like nitrogen, is chemically very closely related to carbon. While nitrogen proceeds carbon on the periodic table, boron precedes it. Boron is a non-metal with an atomic number of 5, and an electron configuration of 2-3. Boron also slots nicely into the carbon lattice, but with a catch. The reduced number of electrons (when compared with carbon) leaves a 'hole', or space in the lattice.

Here the boron atom is called an acceptor, as it is able to accept another electron from an external source. The acceptor fills another, lower, level in the band-gap, awaiting an electron to fill the 'hole'. An electron from the valence level fills the 'hole' and in the process, energy is absorbed, producing the complementary blue colour. This requires a much lower energy level, equating to lower frequencies (hence the blue complementary colour), and this can be achieved by the heat around us at normal room temperatures[6] (see Figure 7.3b).

However, this transition of an electron to fill the 'hole', leaves a residual 'hole' in the valence band from which it came. These new 'holes' in the valence band allow electrons collectively owned by the lattice to move in an electrical field, causing the diamond to be a conductor of electricity.[6] This is a property attributed to many metals, which is why metals are commonly used for the wires of electric circuits.

Both type Ib and type IIb synthetic diamonds, like their natural counterparts, owe their colour and classification to the presence of impurities. Synthetic type IIa diamonds are by definition free from impurities. They are colourless, when natural or synthetic, and are the epitome of pure diamond carbon in all respects. Type IIa diamonds, as would be expected of pure carbon, are almost perfect thermal conductors, but are very rare in nature. With synthetic type IIa diamonds, while subject to the band-gap theory, there are no defect-related localized states in the band gap, and therefore no extra 'holes' or electrons. They are by all accounts (if flawless) the perfect diamond.

Why then are all synthetic diamonds not made as type IIa? The problem in producing colourless diamonds arises from the fact that nitrogen is a very intrusive and abundant gas. It manages to find its way into the mixture, even if it is not added to the feed compound. How then do we produce the ideal type IIa colourless gem-quality synthetic diamonds? Near-colourless synthetic type IIa diamonds require concentrations of no more than a few ppm of nitrogen, and these targets have been achieved.

The chemistry of controlling colour

There a number of techniques used in the production of type IIa and IIb synthetic diamonds. One kind of technique involves the depletion of the nitrogen active in the process. This does not eliminate the nitrogen problem, but does reduce it significantly. Another is used in tandem with the chemical manipulation of the original carbon used in the synthesis to improve the purity of the resulting synthetic diamond. The actual colour of the synthetic diamonds is of little or no importance to the manufacturers, or the end consumers, as these stones are not targeted to the gem-stone and jewellery market, an arena where colour (or lack of colour) is highly important. The importance of colour, and the endeavour to control the resulting colour of synthetic diamonds is symptomatic of the need to control other mechanical and electrical/thermal properties. The visual characteristic of colour, is to this end, a signal of these other properties. Colour can be used as a type of indicator, hinting toward the underlying presence of impurities and their complementary mechanical properties. An example of this is the previously mentioned partnership between boron, blue coloration and electrical conductivity.

As a result of this interdependence, it has been concluded that in order to effectively control the colour of synthetic diamonds (grit or mono-crystals), it is necessary to control the presence of these impurities. Ensuring no impurities will effectively produce colourless synthetic diamonds. Nitrogen will naturally find its way into the solution by itself, without the need of intentional addition. However, extra nitrogen can be added for specified industrial purposes. The addition of boron to produce the above-mentioned blue semi-conducting synthetic diamonds, will also involve steps to simultaneously reduce the presence of nitrogen as well.

As discussed, the particular catalyst used during synthesis may have great bearing on the parameters of the technique. The catalyst also impacts on the resulting synthetic diamond in another very important way. The chemical constituents of the catalyst assist in the addition, or removal, of impurities from the synthetic diamond during crystallization. The most commonly used catalysts are nickel, iron and iron-based alloys, which generally produce yellow synthetic diamonds. These do not contain elements that react chemically with the nitrogen, in favour of carbon.

However, other metals do react favourably with the nitrogen present, and effectively reduce the nitrogen content in the synthetic diamonds. These are termed 'getters', as they basically reduce the opportunities for the nitrogen present to combine with the carbon and therefore infiltrate the synthetic diamond lattice.[1] The most widely used 'getter' is aluminum (Al), however titanium (Ti), and zirconium (Zr) will also be successful in achieving the desired results. 'Getters' are combined with the catalyst in the form of an alloy with the other metals, and during synthesis 'consume' the majority of the nitrogen, creating colourless gems.[2] Without 'getters' the usual amount of nitrogen in synthetic diamond is a concentration of about 200–300 ppm. With a large concentration of 'getters' in the catalyst, the concentrations of nitrogen in synthetic diamond lattices have been reduced to around 20 ppm, resulting in near-colourless gems.[1]

Another technique used to produce pure diamond for industry, that does result in colourless gem-quality synthetic diamond, is to use the purest carbon available. As discussed in Chapter 3 on carbon, this element exists in a number of isotopes. These are pure carbon 12, the heavier carbon 13 (with an extra neutron in the atomic nucleus), carbon 14 (used for carbon dating) and carbon 11. Carbon 12 is the commonest carbon isotope and represents around 98.89 per cent of all carbon atoms;[46] however, the presence of that one-in-a-hundred carbon 13 (or 14 or 11) can have serious implications for high-tech applications of synthetic diamonds. Here the slightest imperfection and disaccord within the crystal could affect the high precision application and hinder applicable technology. For example, using only the isotopically pure carbon 12, synthetic diamonds have been produced with increased thermal conducting properties, making them the best thermal conductors known.[2]

Scientists at the General Electric company were the first to perfect the method of synthesizing isotopically pure diamonds. This means that they were able to achieve diamond synthesis using only one of the carbon isotopes. Researchers at Oak Ridge National Laboratory (among others) have been able to separate the carbon atoms into their respective isotopes. This is an expensive process and the cost of making such diamonds is significantly greater than their 'mixed isotope' counterparts. The GE team first used the isotopically pure carbon 12 in the form of graphite to produce synthetic diamond grit. This grit was then used as the feed material, and converted to a synthetic diamond mono-crystal. Both procedures employed standard techniques. The result is the purest of the pure![2] Again, the near-colourless visual appearance is a by-product of the need to make ultra-pure commercial diamonds.

The new method announced in 1990[47] incorporated isotopically pure diamond powder (formed originally by the CVD, a process outlined in Chapter 14) as the feed material, rather than graphite. This technique can also be used to produce isotopically pure carbon 13 (or ^{13}C) diamonds, with their unique characteristics, for specialized purposes.[46] In 1993, the GIA Gem Trade laboratory at Carlsbad in California, USA, accessed two of the General Electric isotopically pure carbon 12 (or ^{12}C) synthetic diamond crystals for the purposes of gemmological analysis. These synthetic diamonds were not only produced from 99.99 per cent ^{12}C, but were produced using getters in the catalyst, and were therefore relatively pure and free from colour.[46]

The GIA GTL studies indicated that these synthetic diamonds had no bands in the visible spectrum, no metallic inclusions (discussed in Chapter 10), nor fluorescence. All of these characteristics were observed previously by the GIA in other near-colourless synthetic diamonds tested by the GIA. The samples were however suprisingly free from internal strain.[46] These gemmological properties are explained in more detail in the next few chapters. Colour is not, however, the only area in which the chemistry of diamond synthesis can exert some jurisdiction, nor is it the only feature of importance in the study of synthetic diamond crystals.

Morphology

The morphology of synthetic diamonds varies considerably between three of the ideal platonic states. These are octahedron and cube, and a variation on the platonic dodecahedron – the rhombic dodecahedron.[4] All three of these structures occur in natural diamonds, though natural and synthetic crystals look entirely different. To begin we will discuss the shapes, show how these can be altered chemically and then make an overall comparison of the morphology of synthetic diamond mono-crystals with that of their natural counterparts.

Although tending to the perfect forms of the platonic solids (cube, octahedron and dodecahedron), in the field of diamond synthesis these are an almost unobtainable ideal, a virtual nirvana. Only in the rarest of occasions, and under strictly controlled circumstances (as we will discuss in the next section) will synthetic diamond mono-crystals grow into these perfect geometric forms.

The usual growth habits for synthetic diamond mono-crystals are variations and combinations on the structures introduced above. The preferred habit for the synthetic diamond, the most commonly occurring (and most economically feasible) is a series of cubic to octahedral forms.[6] The term then applied is cubo-octahedral or cuboctahedron, covering all possibilities regardless of dominance in either the cubic {100} or octahedral {111} forms. The rhombic dodecahedral {110}, and other more complex forms, manifest (in particular) in crystals of very high purity. Therefore, the rhombic dodecahedral shape is seen only in colourless crystals.[6] Additional unusual shapes have been reported, notably a General Electric crystal that formed a rare 'fiveling' twin in the shape of a five-pointed star. It was complete with twin planes at the four re-entrant angles, in the octahedral direction [111], with the fifth re-entrant being a mismatched meeting rather than a true twin.[6]

The variations in morphology can also be seen in synthetic diamond grit, and are not necessarily unique to synthetic diamond mono-crystals. Indeed the morphology of synthetic diamond grit shows a greater level of variations (than mono-crystals) and combination forms. The controlling of the shapes and morphology of synthetic diamonds plays a large part in the manufacture of different products, and will be described in the next section.

One point that the graphical representation of the combination of the platonic solids, manifesting in synthetic diamonds, fails to depict is the continuous series of states that may occur. The fact that any combination may be produced rather than the examples of the combined forms depicted in Figure 7.4, is a difficult concept to present within the confines of this text; however with some imagination we can see the transgression.

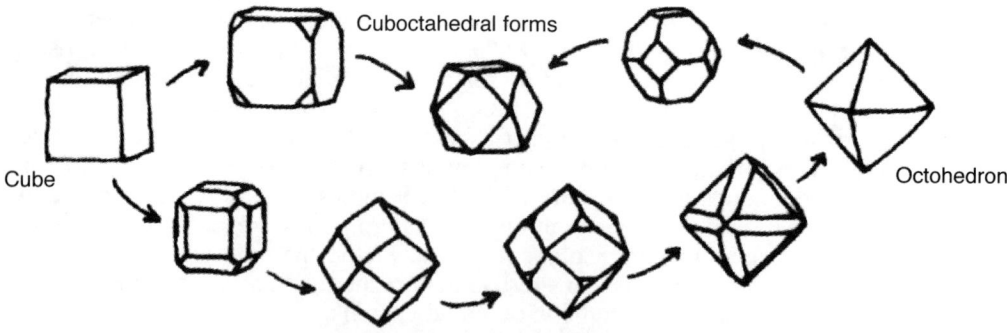

Figure 7.4 Schematic diagram representing common morphological habits of synthetic diamond crystals. Arrows represent regions of mixed morphology and variations of shapes shown.

To date the GIA Gem Testing Laboratory have reported testing a total of 22 colourless and near-colourless synthetic diamond mono-crystals provided by a number of manufacturers as follows:

- General Electric Company (five synthetic diamond mono-crystals)
- Sumitomo Electric Industries (three crystals)
- Russian synthetic diamond crystals distributed by Chatham Created Gems (six crystals)
- De Beers of South Africa (six crystals)
- total of two of unidentified manufacture (supplied by Starcorp Incorporated).

It is important to note that the majority of these crystals were experimental products, rather than commercial products, and the data was compiled from testing performed over the period between 1984 to 1997.[46]

All of the samples were determined to be type IIa, through testing of their spectra and infrared spectrum, although a few were found to be combinations of types IIa, IIb and IaB.[46] These 22 near-colourless synthetic diamond crystals were described collectively as having 'a very distinct shape that consisted of a portion of a cuboctahedron with a flat base'.[46] The report stated that cubic (100) and octahedral (111) faces were the most prominent on the samples, accounting for the largest surface area. Additionally the smaller faces were present in the dodecahedral (110), trapezohedral (113), and (115) directions.[46]

To assist in visualization of this myriad of faces and miller indices, consider the idealized crystal in Figure 7.5. Here the faces have been denoted with letters (rather than index numbers) and their shape enhanced. The cubic faces are denoted by the letter 'c', the dodecahedral by 'd', the octahedral by 'o', and the trapezohedral faces are those indicated by 't'. Actual crystals, however, are not as crisp and precise as this, and their faces are less regular and uniform.

The description in the GIA report also summarized the nature of the faces being flat, with sharp edges and corners; and defined the basal face as being a growth surface, rather than a true crystal face. The report also made mention of the absence of the mechanical and chemical etching that is present on natural diamond crystals, and instead exhibited a type of surface marking of their own.[46] This comparison is included in a later section.

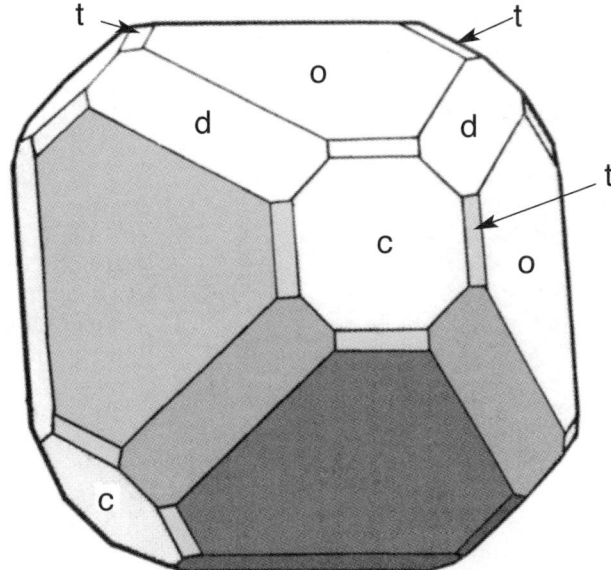

Figure 7.5 Idealized diagram of synthetic diamond crystal in a modified cuboctahedral form. Faces identified with c = cubic, o = octahedral, d = dodecahedral, and t = trapezohedral. (Courtesy *Gems and Gemology* **23**, 4 © GIA 1998.)

These morphological features are commonly reported in synthetic diamond monocrystals. Two Sumitomo yellow synthetic diamonds were also tested by the GIA, and the results published in 1992. The first 5.06 ct sample, although polished on two sides, parallel to the cube direction (100) was reported as having been predominantly cubic in shape, with evidence remaining of cubic (100), octahedral (111), dodecahedral (110), trapezohedral (113) and (115) respectively, present to varying degrees. The second sample being 5.09 ct, although prepared in the same way, was described as exhibiting larger octahedral (111) faces, smaller cubic (100), trapezohedral (113), (115) and dodecahedral (110) faces respectively.[38] Here also striations and surface patterning were reported.

Synthetic diamonds occur in a combination of cubic and octahedral forms. How does this assist us? There are two areas where knowledge of the characteristic shapes of synthetic diamonds can assist. The first is the identification of synthetic diamonds (to be discussed in Chapters 10 and 11), and, second, in the manipulation of these shapes to produce required products for specialized purposes; to be discussed next.

The chemistry of controlling shape

Unlike the controlling of the actual colour of the stone, the controlling of the shape is of great significance to commercial manufacturers and to the ultimate consumers. Synthetic diamond mono-crystals, as we have seen, are formed in extreme environments. This intemperate exposure manifests itself in the morphology of the resulting crystals (whether mono-crystals or grit).

These conditions are responsible for more than just the outward appearance of the synthetic diamond. They are also the cause of the etching and external patterning

70   The diamond formula

mentioned briefly in the last section (to be outlined in more detail next), and the internal growth structure. It is the direction of growth that is the main contributor to the final shape of the crystal.

The ability to achieve cubic or octahedral habits is widely known. Faster growth in the cubic direction [100] results in the development of an octahedron – and conversely faster growth in the octahedral direction [111] results in a cube. However scientists still cannot predict the exact shape that will form during routine synthesis.[2]

Research has been conducted into this problem and many advances have been made. It has been determined that certain parameters have an effect on the development of growth in synthetic diamonds. These parameters include exact pressure, type of catalyst, form of carbon and most importantly, the temperature. The temperature, it was discovered, has the most significant effect on the morphology of the growing crystals.[6]

It was found that perfect octahedrons form more readily at high temperatures (relatively speaking) and cubes form more readily at lower temperatures. The combinations most commonly seen are formed at the intermediate temperature.[6] At 1300°C the synthetic diamonds form in almost perfect cubic-shapes, at 1400°C to 1550°C octahedral faces emerge to form the familiar cubo-octahedral, and above this octahedral faces predominate.[6] Morphology does however, depend upon a number of other factors; temperature is just one.

The modification of exact pressures applied during the procedure can also have an effect on the resulting shape of the crystal. This can be seen in Figure 7.6, where the Berman-Simon line is plotted with a graphical representation of the corresponding morphological states. Pressure alterations can also induce the formation of additional faces. Pure octahedra are rare, with cubic faces usually present, although small.[41]

Catalysts also play a part in the formation of faces in synthetic diamond morphology. For example, research has shown that synthetic diamonds grown in a pure nickel catalyst form pure cubo-octahedra. The employment of other metals has been found to induce the manifestation of minor dodecahedral (110) and trapezohedral (113) faces. The additional trapezohedral (115) face, mentioned in previous descriptions, is due to the addition of cobalt to the catalyst metals, in combination with the nitrogen 'getters'.[41]

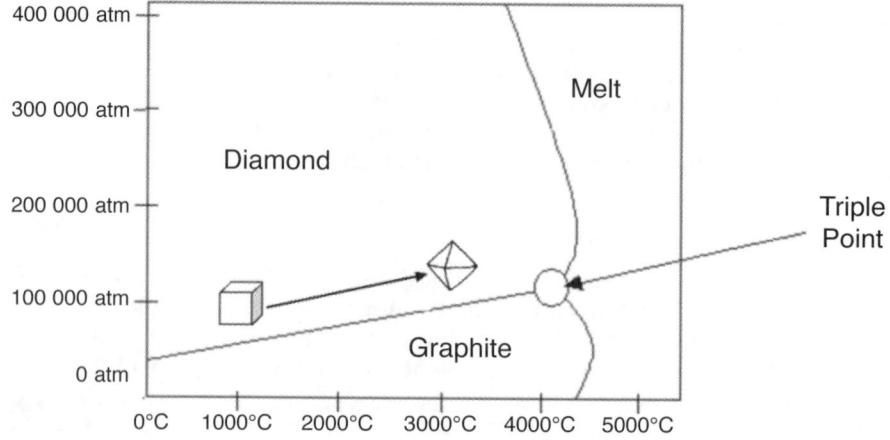

Figure 7.6 Phase diagram for carbon featuring approximate regions with corresponding morphologies.

Naturally the other parameters mentioned have similar effects as listed above, to a somewhat lesser degree. These are exploited less, and attention is given mainly to the manipulation of temperature and the selection of catalysts in controlling the shape of synthetic diamonds. In all, up to 40 faces have been reported on a single 5 mm crystal of synthetic diamond.[6] This overall shape is in sharp contrast to the morphology of natural diamonds.

The natural comparison

In nature, diamonds grow in a very different way to that of synthetic diamonds. While their environments are similar, both at extreme pressures and temperatures, and in combination with other materials, the way the carbon is deposited is of major contrariety. First we will take a brief look at the morphology of natural diamonds to elucidate our overall comparison of natural and synthetic diamond growth habits.

The morphology of natural diamonds varies from different countries and indeed individual mines. We will generalize here for the sake of impending comparison with the relative uniformity of synthetic diamonds. This fact alone is one component to be considered in the comparison of the habit of natural diamond, and that of their synthetic cousins. Natural diamonds offer a wider range of shapes, and manifestations. The variance in the size and number of faces, overall forms, twinning and facial patterning is remarkable, a veritable history of their development. This is an obvious contrast to the development of synthetic diamonds, and the way their growth is expressed in their external habit.

Natural diamonds grow in reasonably even concentric layers, akin to that of a natural pearl. Once the crystal has nucleated in the volatile magmas deep within in the earth, the path is set. Usually the crystal grows in the same direction throughout its development, without much deviation from the course chosen by the nucleation beginnings. Sometimes natural diamonds do change their mode of growth from octahedral or cuboid (or vice versa).[145] In most cases though, if natural feed solution is denied to a certain area of the crystal, this side will no longer increase in dimension, but the remaining crystal will still maintain growth in the same direction, resulting in the designated crystal form, although a little lop sided. The crystal growth of natural diamonds is uniform, forthright and with the utmost integrity. This can be seen by examining the zoning and graining in natural diamonds, both of which are representations of the growth and ultimately the overall morphology of the natural habit.

The growth of synthetic diamonds is less than uniform. Faces on synthetic diamond crystals, as discussed in the previous section, are representative of growth sections. The cubo-octahedral morphology is a direct result of growth in both the cubic and octahedral directions, simultaneously. Other faces are also a result of differential growth sections in the respective directions. In addition to this the direction of growth of synthetic diamonds may change, merge or splinter, as the crystal increases in size. This is due in part to fluctuations in the temperature, pressure, the distribution of the carbon within the catalyst, and rate of deposition. This is why synthetic diamonds exhibit the combination form of cube and octahedron, and why other faces manifest in a way that seems extraneous in relation to the rest of the crystal.

Natural diamonds form in a number of geometric states. These are purer in form, and are octahedral, cubic, and rhombic dodecahedral and, to a lesser extent, trisoctahedral and hexoctahedral. Octahedral habit, or variations on this theme, are the most common form of natural diamond, and are preferred by diamond cutters. Small natural diamonds have been found in the cubo-octahedral form but this is very rare.[41]

Flat dodecahedral and trapezohedral faces have never been reported in natural diamond crystals. All manifestations of dodecahedral faces are due to dissolution.[41] Twinning is also possible in natural diamonds. Twins are spinel-type contact-rotational twins, with the twin plane parallel to the octahedral direction, and are called macles.[5] The cyclic twinning described previously, accounting for the GE 'fiveling' is not seen in natural diamonds.

Natural diamonds are also characterized by surface markings of various types. The most common and widely known are the triangular pits in the surface of the octahedral faces of the natural crystal. These are termed 'trigons' and have reverse orientation with respect to the crystal face. A similar phenomenon is seen in cubic crystals, with square pits known as 'quadrons' in the surface of the faces, oriented inversely with respect to the face. While rhombic dodecahedral natural diamond crystals do not have uniquely pitted faces, they do exhibit parallel lateral surface striations, oriented along the longer direction on the rhombic faces, mainly due to dissolution.

Natural diamonds exhibit an overall rounding of the crystal. This is not due to wear as was once thought, but to the volatile conditions experienced by the diamond crystal during emplacement. The chemically and temporarily hostile conditions during emplacement cause dissolution of the surface of the crystals, literally eaten away by the kimberlite and lamproite magmas, causing a type of molecular stepping (and dodecahedral faces, with striations). Pronounced stepping, due to dissolution, gives the crystal a grooved or corrugated appearance, and the term 'crinkles' is applied.[5] Natural diamonds often appear with a gum-like skin or coating, called the nyf.[5] It is this coating that can sometimes give rise to the appearance of apparently flat dodecahedral faces in natural crystals.[41]

While there is a type of surface etching present in the synthetic crystals, the appearance is different in many respects. Forms of dissolution and etching are absent from synthetic crystals, however they have a potentially diagnostic surface 'fingerprint' of their own. Surface patterning reported in synthetic diamonds include dendritic or striated markings of a type not seen in natural diamonds. These are due to imprintation by the solidified metallic catalyst.[38] In addition to this, trigons (the triangular pits seen in natural octahedra) have been observed on the surfaces of synthetic diamond monocrystals, dispelling the notion that such markings are always diagnostic of natural origin.[46]

Such knowledge of the morphology of synthetic and natural diamond crystals can assist in the identification and separation of samples. However, once faceted, all of these features are lost. From here we must delve into the optical and gemmological properties of synthetic diamond mono-crystals if we are to reach a higher level of understanding, to eventually enable gemmological separation.

Chapter 8

Gemmology of synthetic diamonds

Soon after the successful synthesis of diamonds in the late 1950s, researchers from General Electric published a paper stating that there were no detectable differences between natural and synthetic diamonds.[48] Thankfully, this is now known not to be entirely accurate. In fact, only weeks later in the same journal, evidence was supplied by another research team of distinct microtopographical differences.[49, 50] Today we have a reasonably large amount of observational data with which to compile a list of differences between natural and synthetic diamonds. Most importantly, many of these differences are easily detectable to suitably trained individuals in the gem industry. The following chapters are devoted to discussion of the characteristics of synthetic diamonds that will eventually lead to a strategy for identification.

Synthetic diamonds share many of their gemmological characteristics with their natural counterparts, both being diamonds and therefore having identical chemistry and molecular structure. We will commence this chapter with a brief discussion of the physical properties of synthetic, and also natural diamonds, the gemmological constants and more complex gemmological properties such as spectra and reaction to ultraviolet radiation.

In the following chapters some attention will be given to more advanced scientific classification of synthetic diamonds, including tests that in most cases are beyond the capabilities of most gemmologists due to instrumental restraints. These results, including cathodoluminescence, are considered important in a full and complete description of the topic at hand. We will then cover inclusions, and related properties, with a short explanation on grading.

Physical characteristics

The standard constants used widely in the description and identification of gemstones can naturally be applied to synthetic diamonds, in the same way that they are applied to natural diamonds. The specific gravity of diamond, discussed in Chapter 3, is 3.515, and indeed, one would expect the same to apply to synthetic diamond when tested in the standard way. However, hydrostatic tests performed by the GIA on samples of yellow Sumitomo synthetic diamonds concluded that the specific gravity differs from that of natural diamonds. In a mixture of Clerici's solution with distilled water (to suspend a natural diamond crystal of SG 3.51) the synthetic diamond rose slowly. Final tests found the specific gravity of these synthetic diamonds to be 3.505.[36] Conversely, synthetic diamonds with prominent metallic inclusions sank.[37] It was decided that minor fluctuations are insignificant in relation to the overall uniformity, and are not sufficient to assist in identification.[36] One might expect to discover that the metallic inclusions (discussed later) frequently found in samples, and often dispersed throughout the synthetic crystal, have a minor effect on the overall specific gravity of synthetic diamonds when routine tests are performed.

The hardness of synthetic diamond is 10 on the Mohs scale, in accordance with the hardness of natural diamond, and the presence of nitrogen has the added effect of making the crystal slightly harder.[6] It is for this reason that no moves are made to eliminate the nitrogen in synthetic diamond grit or industrial yellow synthetic diamond mono-crystals used for cutting, drilling, and grinding. In some cases the synthetic diamond is doped with additional nitrogen, though mainly in an attempt to create n-type conductors.[1]

A type of directionally differential hardness characterized in natural diamonds is used by cutters to enable effective sawing and polishing. A similar phenomenon has been characterized in synthetic diamonds following the statement by the Belgian diamond cutter M. Bonroy that the Russian synthetic diamonds provided for cutting in 1967 were difficult to cut.[5] GIA researchers in 1986 arranged for the faceting of a number of sample synthetic diamonds supplied by Sumitomo, in order to analyse their faceting behaviour. Cutters at the diamond cutting facilities, in New York and Los Angeles, were supplied with nine pieces of synthetic diamond, ranging from 3.55 × 3.54 × 1.63 mm to 4.00 × 3.76 × 1.62 mm in size, in the form of the rectangular tabular pieces prepared by Sumitomo. This is the standard preparation offered to their industrial clients. From these pieces one round brilliant cut stone was prepared, with a maximum yield of 22 per cent of the original piece. The remainder were fashioned into emerald cut stones, with maximum yield varying from 49 per cent to 64 per cent.[36]

The cutters subsequently reported a number of interesting features of the synthetic diamonds, and later confirmed that had they not been told of the synthetic origin of the diamonds, they would still have suspected something unusual about the stones. These features include differences in the ability of the synthetic diamond to cleave and to be sawn, freedom from knots or other potentially problematic defects and the fact that the facets on the synthetic diamonds had only one polishing direction. In addition to this the cutters found that these diamonds were less brittle than natural diamonds. The synthetic diamonds also exhibited mechanical differences too. The cutters found they did not require the same final polishing with the finer grained wheel, and they did not become as hot on the cutting wheel. When increased pressure was applied to the polishing dop (an action that will result in natural diamonds being polished more quickly) the synthetic diamonds removed the diamond powder from the wheel, rendering it useless.[36]

One of the cutters from Lazare Kaplan International had previously faceted some of the early General Electric synthetic diamonds and commented that the Sumitomo characteristics were consistent with these early GE stones.[36]

Optical constants

Gemmological optical features all rely on the interaction of the gem with light (and other forms of electromagnetic radiation such as ultraviolet light and X-rays). The colours of the visible light spectrum were first observed by Isaac Newton in 1666. In 1814 Joseph Fraunhofer observed the spectrum of the light emitted by the sun, and noticed a multitude of absorption lines.[4] These can be easily seen by using a hand held spectroscope of the type used in gemmological inspection of spectra. Fraunhofer labelled the strongest of these lines A to H. These lines can be found at 760.6 nm (A),

687.0 nm (B), 656.3 nm (C), 589.3 nm (D), 527.0 nm (E), 481.6 nm (F), 430.8 nm (G), and 396.9 nm (H) respectively.[51] They were later found to correspond to the spectra of certain elements, for example it was ten years later that Gustav Kirchhoff established that the element sodium was responsible for the D line.[4] The significance to gemmology is that the dispersion of gemstones is measured between the B and G Fraunhofer lines.[5]

Dispersion of gemstones is a measure of the level to which the material separates the colours of the visible spectrum, or 'disperses' them. Dispersion is dependent on optical density, and the wavelengths of light to which the sample is exposed.[13] Each wavelength has a varying refractive index, which produces the separation of colours. The dispersion of diamond is measured at 0.044, between the B–G interval.[5] Synthetic diamond is understandably the same.

The refractive index is another important gemmological constant used in the identification and classification of gemstones. The refractive index for diamond is measured at 2.417, for yellow sodium light of 589.3 nm (corresponding to the 'D' Fraunhofer line in the solar spectrum), the universal standard used to measure gemmological refractive index.[51] The respective refractive indexes for diamond at each of the spectral colours are 2.407 for red at 687.0 nm, 2.417, for yellow at 589.3 nm, 2.427 for green at 527.0 nm, and 2.465 for violet at 397.0 nm.[5]

The refractive index of synthetic diamonds is naturally the same as natural diamond. As expected, the path of a beam of light passing through the tetrahedral carbon lattice of synthetic diamond is refracted (due to the differential velocity of the light in the medium compared to air), the same amount from the normal of the plane of entry as in natural diamond. One additional feature attributed to synthetic diamond is that, unlike other gemstones and natural diamonds, the morphology of synthetic diamonds, with their relatively smooth crystal faces, can allow for the testing of refractive index without the need of faceting.

Visual characteristics of synthetic diamonds are also consistent with those of natural diamonds. The unique adamantine lustre, due to the extreme hardness and high density, is identical to that of natural diamonds, when cut. Faceted synthetic diamonds also produce the same fire and brilliance ascribed to natural diamonds when cut to correct proportions, and are visually the same in all respects. This unfortunately concludes that visual identification for the purpose of separation of synthetic diamonds from natural diamonds is not possible, at least not in the way diamond can be visually separated from many of its simulants. Cubic zirconia, for example, allows visual confirmation via optical characteristics like the renowned 'see-through' or 'dot-ring' effect.[51]

Optical characteristics

As we have seen however, many of the morphological features of synthetic diamonds differ from those of natural diamonds, and this results in unique gemmological properties and phenomena. The first faceting-quality, near-colourless synthetic diamond mono-crystal to be described gemmologically was a General Electric specimen tested in 1971 by Robert Crowningshield.[46] From that time a reasonable amount of data has been collected and general descriptions compiled. The reason for the limitations on gemmological data is the fact that the currently produced synthetic diamond monocrystals are produced only on a small scale, and are for experimental and commercial use only (with the exception of a few Russian sources). We will now continue on from the features described in the previous section, using these and other data to form a gemmological portrait of synthetic diamonds, either faceted or raw crystals.

The next optical feature is birefringence. Diamonds, crystallizing in the isometric (cubic) system, by definition, do not exhibit birefringence as the structure is singularly refractive. There is no plane polarization of the light, and no pleochroism.[51] However, synthetic diamonds, formed in extreme environments, often have residual strain in their lattice. Strain can also be seen in some natural diamonds.[51] This strain causes a weak anomalous double refraction (ADR).

ADR is present in the case of both near-colourless and coloured (yellow and blue) synthetic diamonds. This weak strain can also be identified by the presence of low order interference figures, usually seen as black, grey and white cross-shaped patterning between crossed polars.[46] The pattern was observed in the yellow Sumitomo synthetic diamond pieces, analysed in 1986; however, it was not visible once faceted.[36]

Absorption spectra, another valuable tool in the testing of gemstones, is used in the testing of diamonds to assist in classification. The spectrum of diamond is due to the crystalline structure, or faults in the structure.[5] Natural diamonds exhibit diagnostic absorption bands,[51] and these will be compared with synthetic diamonds in Chapter 12.

GIA reported that in the vast majority of the 51 near-colourless synthetic diamonds tested by their laboratory, no significant spectra were observed. Tests were performed with both the hand-held gemmological spectroscope and a spectrophotometer.[46] Synthetic diamonds display an increasing, general type of absorption toward the violet ends of the spectrum,[51] and the GIA have reports of a well-defined band at 270 nm in near-colourless synthetic diamonds. Weak bands at around 732 nm recorded in earlier specimens provided to the GIA (General Electric 1984), were not seen in subsequent samples, and are therefore not considered by the GIA Gem Testing Laboratory to be of any great importance.[46] Their inclusion here is primarily for the purposes of accuracy, and not as a readily observable feature.

In Figure 8.1 we can see the general absorption towards the violet (400 nm) end of the visable range of the electromagnetic spectrum as a sharp incline at the left side of the graph. This data was collected with a Pye-Unicam 8 800 UV-VIS spectrophotometer of crystals supplied by De Beers. The three colours, yellow, greenish and brownish, all show similar results, however the greenish stone exhibits some absorption towards the red end.

The spectral features, or more accurately the lack of diagnostic spectra, come as no surprise. As we have discussed in Chapter 7, synthetic diamonds are type Ib, IIa and occasionally IIb (the latter by chemical manipulation). These types of diamonds do not usually exhibit diagnostic spectral phenomena in the visible part of the spectrum.[52]

A number of other spectral bands have been seen in synthetic diamond, attributed to nickel used as a catalyst, included within the diamond, and dispersed through the crystal. These lines (too thin to be classed as bands) appear between 470–700 nm, and observation of such lines may be used for the purposes of identification.[52] A listing of these lines is provided in Table 8.1, showing those lines visible with the hand-held gemmological spectroscope. These observations were taken from testing two treated synthetic diamonds (outlined in Chapter 9), and were recorded at liquid nitrogen temperatures.[53]

The yellow coloured Sumitomo synthetic diamonds showed a similar reaction to General Electric synthetic yellow diamonds, with no spectral patterns of absorption or emission. This was confirmed in the case of the Sumitomo synthetic diamonds,

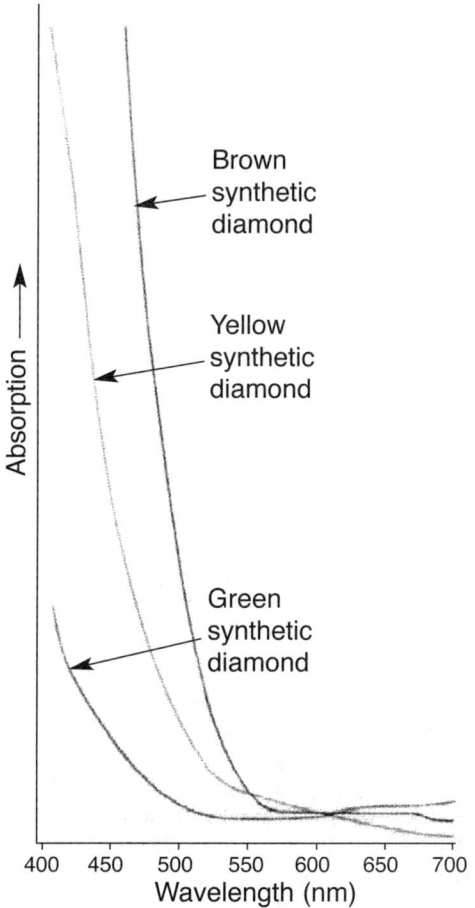

Figure 8.1 Visible range absorption spectra for three colours of De Beers synthetic diamonds obtained using Pye-Unicam 8800 ultraviolet-visible spectrophotometer at room temperature. (Courtesy *Gems and Gemology* **23**, 4 © GIA 1998.)

tested at 60K with a Pye-Unicam dual beam ultraviolet-visible spectrophotometer.[36]

General Electric synthetic diamonds were found to exhibit no spectral bands at all. All colours were tested and the testing extended to include cryogenic (low temperature) examination with negative results. One General Electric synthetic diamond did show a faint line at 415.5 nm (diagnostic for the Cape spectrum of natural diamonds, as we will determine in Chapter 12); however, this was attributed to the small natural diamond seed inserted to induce formation.[37]

We now move from the visible to the ultraviolet arena. So far many of the features we have seen have been nothing spectacular. The results of previous observations have been akin to those expected by study of natural diamonds, or these have yielded nothing spectacular at all. The reaction of synthetic diamonds to ultraviolet radiation represents a rapid departure from this situation.

Table 8.1 Visibility of spectra lines in synthetic diamond due to nickel and nitrogen.

Wavelength (rounded to nearest nanometre)	Chemical relationship (nickel/nitrogen)	Visible with hand-held spectroscope
792 nm	Nickel related	No
732 nm	Nickel and nitrogen related	No
711 nm	Nickel and nitrogen related	No
691 nm	Nickel and nitrogen related	No
658 nm	Nickel related	Yes
553 nm	Nickel and nitrogen related	Yes
547 nm	Nickel and nitrogen related	Yes
540 nm	Nickel and nitrogen related	Yes
527 nm	Nickel and nitrogen related	No
520 nm	Nickel and nitrogen related	No
518 nm	Nickel and nitrogen related	Yes
516 nm	Nickel and nitrogen related	No
511 nm	Nickel and nitrogen related	Yes
503 nm*	Nickel and nitrogen related	No
502 nm	Nickel and nitrogen related	No
494 nm	Nickel related	No
491 nm	Nickel and nitrogen related	No
478 nm	Nickel and nitrogen related	No
473 nm	Nickel and nitrogen related	No
468 nm	Nickel and nitrogen related	No

*Also related to H3 centre in treated pink diamonds

Ultraviolet radiation, or UV light, is used in gemmology to assist in the identification of most species. It is achieved by irradiating the specimen with certain wavelengths of light corresponding to those in the ultraviolet region of the electromagnetic spectrum (400–200 nm). To introduce a level of consistency in the use of UV, two such wavelengths have been established as standards. These are taken as lines in the mercury spectrum and correspond to 365.0 nm and 253.7 nm. These wavelengths have been given the names long-wave (LWUV) for 365.0 nm, and short-wave (SWUV) for the sake of convenience.[51] The reaction to this radiation is measured as the emission of visible light by the sample under test, and is called fluorescence.[51]

Almost all diamonds, whether natural or synthetic, are stimulated in some way by ultraviolet radiation. Usually, synthetic diamonds respond to short-wave UV, and not to long-wave. For example, of the 51 synthetic near-colourless diamonds tested by GIA, only two have shown reactions to long-wave ultraviolet light, and only one such sample failed to react to the stimulation of short-wave ultraviolet. This particular synthetic diamond was of unknown origin (supplied by Starcorp). One of the samples showing long-wave UV fluorescence was of unknown origin (supplied by Starcorp) and the other was from Sumitomo.[46]

In the case of near-colourless synthetic diamonds and most yellow synthetic diamonds, it is believed that reactions observed under long-wave UV will be superseded by the stronger, more pronounced reaction visible under short-wave UV.[52] Reactions to short-wave UV vary among near-colourless synthetic diamonds, but can be generally described as moderate to strong and of generally yellow in colour. Variation in colour ranges from yellow, whitish-yellow, orange-yellow, orange, yellow-green, greenish-yellow, green-yellow and green.[46]

The General Electric near-colourless synthetic diamonds were found to fluoresce very strongly to short-wave ultraviolet light, with a yellow colour and patterning. The greyish-blue synthetic diamonds produced by General Electric exhibited a very strong greenish-yellow fluorescence and yellow synthetic diamonds curiously displayed a different result, being inert to short-wave ultraviolet stimulation.[37]

Yellow coloured type Ib synthetic diamonds from Sumitomo have also proved to be inert to long-wave UV.[38] De Beers[54] and Sumitomo[36] yellow synthetic diamonds all were inert to long-wave UV. The De Beers yellow synthetic diamonds were found to exhibit strong yellow and greenish-yellow fluorescence when as-grown; however, faceted specimens were either inert to short-wave UV or fluoresced with a weak yellow.[54] Sumitomo yellow synthetic diamonds displayed a weak orangy-yellow,[38] and moderate to strong yellow or green fluorescence with the core of the crystal showing a distinct green.[36]

Yellow synthetic diamonds from Russia (loaned to the GIA by Chatham Created Gems) have displayed a rather more unusual result to ultraviolet radiation. First, it was found that the fluorescence varied in intensity with regard to the orientation. Additionally all samples displayed some relatively less intense fluorescence to long-wave UV, and a number of treated samples annealed by high-pressure heat treatment following synthesis, displayed the reverse results, with the stronger results visible under long-wave UV, rather than short-wave UV.[33] Other Russian synthetic diamonds have yielded similar results,[55] including those procured and tested by the author of this text.

The fluorescent colours may result from the presence of impurities, either nitrogen or catalyst metals, as they have the propensity to incorporate exclusively into certain growth sectors that then exhibit fluorescent colours. Nickel-grown, nitrogen-reduced synthetic diamonds show a green colour; while heat-treated, cobalt-grown synthetic diamonds show a yellow fluorescence.[41]

Fluorescence in synthetic diamonds displays another characteristic feature. It represents another manifestation of the internal growth structure of the synthetic diamond mono-crystal. Studies of the fluorescent behaviour of all types (near-colourless and yellow coloured) of synthetic diamonds have reported differential coloration, and variable intensity between the growth sectors of the samples.

Fluorescence is usually unevenly distributed with evidence of growth sectors either in the form of black (lack of luminescence)[46] or fluorescent patterns. These patterns may be cross-shaped,[46] octagonal,[52] or with pronounced banding.[41] Indeed in the Russian yellow synthetic diamonds tested by GIA in 1993, the luminescence was observed only in certain growth sectors or at the intersection of sectors.[36] For example, two Sumitomo yellow synthetic gem-quality diamonds were tested by GIA in 1992. One exhibited weak orangy-yellow fluorescence in the cube sectors but was inert in other sectors. The other had an inert centre with yellow fluorescent outer sectors (parallel to octahedral faces).[38]

Exceptions to these universally observed variable patterns are two isotopically pure type IIa near-colourless General Electric synthetic diamonds tested by GIA in 1993. Here no fluorescent pattern was observed at all.[53]

Phosphorescence is another property exhibited by diamonds, and is closely related to fluorescence. Phosphorescence is defined as the luminescent glow persisting after the exciting radiation has been removed.[51] Many gems exhibit phosphorescence, natural diamonds and synthetic diamonds among them. Categorical reports of the near-colourless synthetic diamond collection tested at the GIA laboratory found that most samples continued to exhibit phosphorescence after the ultraviolet light was shut off. They described the phenomena as weak to strong in intensity, and yellow, yellow-green or blue in colour. Some samples phosphoresced so strongly that samples could be easily seen from a couple of metres in a darkened room. The duration of the observed phosphorescence ranged from 15 seconds to 60 seconds,[46] up to two minutes.[53] Of the coloured General Electric synthetic diamonds, the near-colourless and the yellow stones showed a persistent and strong result, but the greyish-blue type IIb synthetic diamonds did not.[37] The conclusion is that that type IIb synthetic diamonds do not show phosphorescence.[37]

This leads us to the other more advanced forms of luminescence observed using specialized instrumentation beyond the capabilities of the average gemmologist. These are contained within Chapter 9, with additional information on other specialized techniques.

Chapter 9

Scientific testing of synthetic diamonds

In this chapter we will discuss six of the more advanced techniques for the analysis of solid state materials, and in particular, their application to synthetic diamond samples. As previously stated, in most cases these techniques are beyond the capabilities for the average gemmologist. More easily accessible techniques for accurate verification of synthetic diamonds, and separation from natural diamonds, will be contained in the next chapters. Although many of us will not have the opportunity to apply the tests outlined here, it is an important part of the overall classification and description of the subject.

The techniques described here are, for the most part, more advanced derivations of the gemmological tests described in Chapter 8. The first five are based on the interaction of the sample with radiation (of differing wavelengths and frequencies). In the same way gemmological techniques observe the ultraviolet fluorescence and absorption spectra of visible light. These scientific techniques take testing to the next level. The sixth is related to the determination of the chemical composition of the material. We will first examine luminescence, beginning with the best known example in the field of synthetic diamonds, cathodoluminescence, followed by the lesser know phenomena and techniques.

Cathodoluminescence

As the name suggests, cathodoluminescence is a form of luminescence similar to that emitted by exposure to ultraviolet radiation (or fluorescence) described in Chapter 8.[5] In many ways the two are related, for both emit a visible glow, and are representative of the internal growth structure. Cathodoluminescence is produced in a specialized instrument known as a luminoscope.[52]

The luminoscope is manufactured by Premier American Technologies Corporation and is reasonably expensive at around US$20 000.[52] It comprises two parts, including a suspended stage on which to place the sample synthetic diamond. This stage can be sealed to form a chamber, which is then evacuated to form a stable vacuum. The evacuated chamber is attached to a gun-like apparatus that emits the stimulating radiation. The second part of the luminoscope is the unit designed to control the power and intensity of the emission.[52] For those with access to such equipment the luminoscope proves easy to use, and has become standard at gem-testing laboratories.[46]

Cathodoluminescence is the emission produced when a sample is exposed to cathode rays. While ultraviolet light is electromagnetic radiation of shorter wavelengths (than those of visible light), cathode rays are composed of a stream of electrons. Cathode rays were first generated by means of the Crookes tube, an invention of the British physicist Sir William Crookes.[56] It was H. Michel and G. Riedl, at the firm G.L. Hertz in Vienna, who in 1920 first produced readily available apparatus for the production of cathode rays. This device consisted of a glass bulb containing three electrodes, in which the sample was placed. The bulb was evacuated by a mercury pump, and the arrangement of the electrodes could be changed to produce either cathode rays or X-rays.[5] Cathode rays were first applied to the study of gemstones by H. Michel and W. Crooke.[5] Their use in the study of diamonds has proved its value with the advent of synthetic diamonds.

Cathodoluminescence is generally considered to be the most effective and conclusive test to identify synthetic diamonds.[46] There are two main reasons for this. First is the fact that no diamond, either synthetic or natural, is inert to cathodoluminescence,[52] and, second, it produces clear and diagnostic results.[52]

The cathodoluminescence exhibited by synthetic diamond differs greatly from that of natural diamond. The multi-coloured emissions of both materials indicate the growth patterns and speed at which the crystal formed. Luminescent bands, lines, and sectors of varying degrees provide a veritable fingerprint for their creation. Either their even concentricity or geometric arrangement tell the history of the stone. The cathodoluminescent pattern exhibited in synthetic diamonds is very similar to that observed under ultraviolet light. Both are a geometric configuration of luminescent growth sectors reminiscent of the internal growth structure of the mono-crystal.[57]

Cathodoluminescence was observed in 19 of the 51 near-colourless synthetic diamonds tested by the GIA. These represented all of the production sources and suppliers for the collection. The uneven pattern was collectively described as cross-shaped, or zoned in appearance, and the colour of the cathodoluminescent emission was observed as blue, green, or yellow green in colour.[46] Similar studies of yellow synthetic diamonds from Russia and De Beers found that the colours of the zoned cathodoluminescence in these samples was orangy-yellow and greenish-yellow,[54] and the pattern a similar geometric arrangement of zones.[33]

In contrast, natural diamonds do not have the same angular arrangement of growth zones, and therefore do not show the uneven arrangement of cathodoluminescent patterns. Natural diamonds exhibit a concentric layered appearance reminiscent of their slow and even layered growth, in the same direction throughout the crystal. This cathodoluminescence is parallel and relatively uniform.[54]

The cathodoluminescence of synthetic diamond allows easy examination of the internal growth structures. It is considered to be brighter, clearer and more distinct than other forms of luminescence.[33] Recent developments in the availability of specialized equipment have made the use of cathodoluminescence less crucial, but no less effective.[46]

X-ray illumination

Unlike cathode rays, X-rays are a very widely known form of radiation. Also unlike cathode rays, X-rays have a number of everyday uses, many of which are familiar. You can observe X-rays at work when you pass your hand luggage through the detectors at the airport, and there is a high possibility that if you have not already, you will at some stage have a medical or dental X-ray photograph taken. In gemmology X-rays are used to separate cultured pearls from natural pearls, and to observe the characteristic luminescence of various gem materials to aid in identification.[5]

X-rays are a form of electromagnetic radiation, similar in nature to visible light but of a wavelength of around 0.2 nm.[51] This is much less that ultraviolet light. Those X-rays closer to ultraviolet light are called Grenz rays.[5]

X-rays were discovered accidentally by German physicist Wilhelm K. Röntgen in 1895[5] and are also called Röntgen rays in his honour.[51] As we have seen above in the case of cathode rays, X-rays can be manufactured using the dual-purpose bulb made by G.L. Hertz. Cathode rays and X-rays are closely related in this way. When cathode rays (fed through a highly exhausted tube carrying a high-tension current and a potential greater than 30 000 volts) strike a surface, X-rays are produced by the destruction of the cathode rays.[5] Modern X-ray tubes have a special target included, orientated perpendicular to the cathode ray (rather than at 45°, as was the case in early models).[5]

X-rays interact differently depending upon the material under test. Some may allow the X-rays to pass through unimpeded and other substances are opaque to X-rays and act as a reflective shield. This is termed the level of transparency. The transparency of a substance to X-rays is related to the atomic weights of the atoms that constitute the lattice. The higher the atomic weight the less transparent is the material. This is why lead (Pb) with an atomic weight of 207.2,[56] is used as an X-ray shield. As diamond is pure carbon (atomic weight 12), it is transparent to X-rays.[5]

Characteristically the fluorescence of natural diamonds to X-rays is white, yellow, green or blue.[5] Intensity may vary within certain parameters; however, it is far more uniform in nature than the fluorescence seen as a result of ultraviolet light.[51] The majority of natural diamonds exhibit a chalky blue colour of moderate intensity, with little or no appreciable phosphorescence, even if one is noticeable following ultraviolet irradiation.[51] For the most part however, natural diamonds show a relatively predictable and consistent result to X-rays, of variable colour but with high uniformity.[51]

Synthetic diamonds show a more complex result to X-ray stimulation. Different results have been recorded for specimens from various sources. We will discuss each case on a source-by-source basis and then collate the results.

First, the Russian yellow synthetic diamonds produced by the BARS system were exposed to X-rays (80 keV, 40 mA) by the GIA gem trading laboratory. A total of 10 samples were tested and the results collectively described as exhibiting little luminescence when compared to a natural type IaA yellow diamond. It was noted that this result is considered consistent with that recorded for yellow General Electric synthetic diamonds tested up to that time, i.e. 1993.[33]

In contrast the De Beers yellow gem-quality synthetic diamonds were tested by the GIA research team with a conventional X-ray unit (72 keV, 13 mA) for a period of approximately several seconds. The specimens tested included brownish-yellow, yellow and greenish-yellow samples. The brownish-yellow samples exhibited a dark-yellow or greenish-yellow luminescence of moderate intensity to the X-rays. The luminescence appeared in a zonal pattern, but showed little to no phosphorescence. The yellow samples fluoresced in a similar manner, showing a zonal yellow or yellowish-green pattern, but of greater intensity. Again there was little or no phosphorescence. The greenish-yellow samples were again consistent with the results for the other specimens, with familiar yellow coloured zonal patterns, but with even more intensity. In addition, the greenish-yellow samples broke with tradition and exhibited strong phosphorescence for a period of more than 10 seconds.[54]

The results recorded for Sumitomo yellow synthetic diamonds differ from those previously mentioned. Sumitomo samples were exposed to an X-ray unit operating at 66 keV and 35 mA. Under these conditions all samples exhibited a bluish-white glow of variable intensity, and characteristically no phosphorescence.[36] This result has more in common with natural diamonds than other synthetic diamonds tested.

General Electric synthetic diamonds have been found to be inert to X-rays,[54] but recent samples of the isotopically pure, near-colourless General Electric samples have proved the exception to this rule. When these samples were exposed to X-rays in 1993, a yellow luminescence was recorded. This in itself appeared consistent with results from Russia and De Beers. However another striking phenomenon was observed. The samples showed strong yellow phosphorescence in both crystals tested. One crystal phosphoresced for a period of around 2 minutes and the other for approximately 10 minutes.[53] This is certainly unusual when compared with the results recorded for all other synthetic diamonds tested over the years.

In general synthetic diamonds characteristically show a yellowish fluorescence to X-rays, with moderate intensity and no phosphorescence, with the exception of the Sumitomo yellow synthetic diamonds with their bluish-white results.

Electroluminescence

Electroluminescence is a lesser known phenomenon appearing as a fluorescent glow in electrically conductive materials when subjected to sufficient electrical stimulation. The conductometer used to test the electrical conductivity of diamonds was applied to synthetic diamond samples and the phenomenon of electroluminescence was observed.

In the majority of synthetic diamonds tested in this way, no electrically conductive behaviour was recorded, as one would expect from type Ib[36] and type IIa[53] diamonds. Anomalous results were found in greenish-yellow De Beers synthetic diamonds, but were attributed to impurities or structural defects.[54]

However, twenty of the near-colourless synthetic diamonds tested by the GIA proved to be slightly electrically conductive, depending upon the point at which the conductometer was placed in contact with the sample. Of these electrically conductive specimens, six synthetic diamonds exhibited electroluminescence. The luminescence observed in these samples was described as sporadic and weak, and of an interesting blue colour, naturally coinciding with the application of the conductometer probe. In addition to this curious result, the electroluminescence appeared to be localized between the probes, and not emitted by the whole sample. This is similar to the electroluminescence observed in many natural diamonds.[46]

One particular Sumitomo near-colourless synthetic diamond crystal exhibited both blue and red electroluminescence, depending upon the point at which the probe was positioned on the crystal. As if this was not peculiar enough, this sample also continued to phosphoresce for a few seconds after the probes were removed.[46]

As few synthetic diamonds are electrically conductive, and the majority of results observed were similar to those seen in natural diamonds, the GIA reported that the results described above were not suitable for the purpose of identification.[46]

Thermoluminescence

As the name suggests thermoluminescence is another type of luminescent glow, in this case stimulated by heat. Thermoluminescence is distinct from the property of thermal conductivity. As we have seen, diamonds are excellent there. Therefore as expected, all of the synthetic diamonds tested with a thermal conductivity meter showed high thermal conductivity.[54]

Thermoluminescence was observed under different circumstances, and not in all samples. Five of the near-colourless synthetic diamonds tested by the GIA exhibited thermoluminescence while immersed in hot water following exposure to ultraviolet radiation. Four of the samples showed a blue colour and one an orange colour. These colours continued briefly after the samples were removed from the water, but while they were still warm. The GIA subsequently reported that this thermoluminescent reaction has not been observed in natural diamonds.[46]

Infrared spectroscopy

William Herschel, a German-born British musician and self-taught astronomer, discovered in 1800 that beyond the red end of the visible range is the infrared region of the electromagnetic spectrum. These frequencies are too long for us to detect with our naked eyes, but are shorter in wavelength than microwaves or radio waves. Herschel achieved this feat with the simplest of apparatus, a thermometer. Herschel used the thermometer to measure sunlight dispersed into its colours by a glass prism. This inevitably led to the question – what is infrared radiation?

This is relatively straightforward, as it is a type of electromagnetic radiation, but the question of how it interacts with matter is complex indeed. The infrared section of the electromagnetic spectrum runs from around 800 nanometres to wavelengths of around one millimetre, however scientific work uses the wavelengths from 2 to 16 micrometres in length for standard work.[4] Although invisible to the eye, infrared radiation can be detected as warmth by the skin, as it constitutes around 50 per cent of the sun's emitted energy.[58, 60]

Actually it is the interaction with matter that proves useful in science. Infrared radiation is absorbed and emitted by many kinds of materials. Briefly the absorbed infrared radiation causes both rotational and vibrational changes in molecules to the material. The vibrational energy levels depend on the types of atoms and functional groups within a molecule, and changes correspond to the ways in which the individual atoms (or groups of atoms) vibrate relative to the remainder of the molecule.

How is this information used in classifying these materials? The infrared absorption and emission characteristics yield important information about the size, shape, and chemical bonding of molecules and of atoms (and ions) in the material under test. The infrared energy emitted or absorbed by a given molecule, or substance, indicates a difference in the internal energy states. These correspond to the atomic weights of the atoms in the molecule, and the molecular bonding forces. For this reason, infrared spectroscopy is a powerful tool for determining the internal structure of molecules and substances, or for identifying the amounts of certain atoms within a substance.[4]

It is this feature of infrared spectroscopy that is useful in determining the type of diamonds.[54] An example is given in Figure 9.1. As we know, diamond types are determined by the impurities present in the lattice, and the bonding patterns of these impurities. This information relating to any impurities present in diamond under test will assist us in classifying the diamond. This is achieved using a specialized apparatus, based on a concept similar to the visible range spectroscopy described in the previous chapter. The energy levels of the absorption and emission are displayed as peaks on a type of line graph (or 'spectrum'), corresponding to the atoms and molecules (characterized by their respective vibrations) within the material.[4] These peaks indicate the presence of different elements in the lattice. These may or may not be impurities.

Infrared spectroscopic analysis performed on synthetic diamonds serves to classify each sample into its respective type. This does not directly determine whether a diamond is natural or synthetic and therefore is not used as a tool in identification. For example, infrared spectral analysis of the 51 near-colourless synthetic diamonds

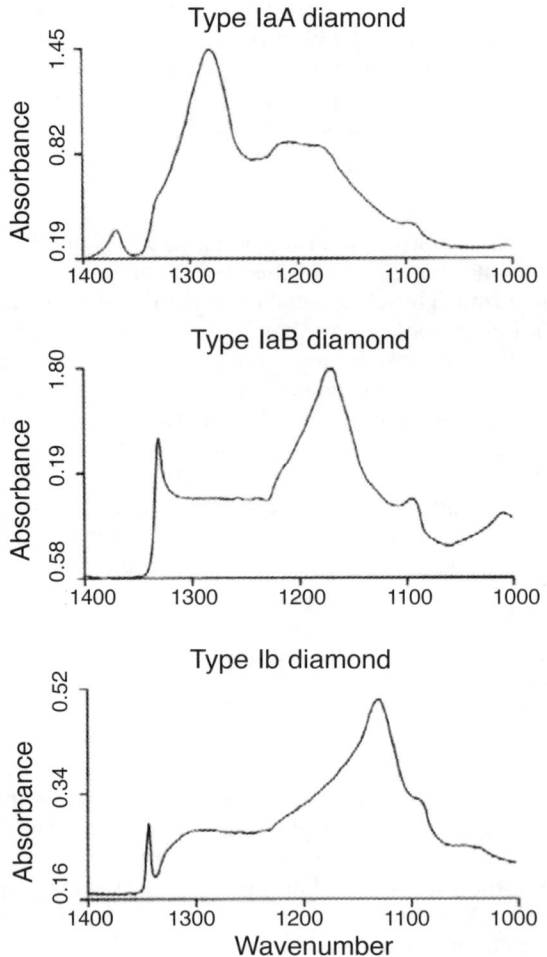

Figure 9.1 Infrared spectra of three diamonds recorded with the GIA's NICOLET 60SX FTIR spectrometer. (Courtesy *Gems and Gemology* **29**, 4 © GIA 1998.)

tested by the GIA was the establishing factor in determining that the type of the samples was in fact IIa.[46] It was also an excellent confirmation in the cases of the two near-colourless isotopically pure General Electric samples.[53]

Study of two yellow gem-quality Sumitomo synthetic diamonds proved very interesting for the GIA in 1992. The infrared spectra of one sample proved that it was indeed a type Ib synthetic diamond, and indicated to the examiners that the crystal was uniform in type throughout. However, the second sample recorded a more unusual result. Different coloured zones had already been identified within the sample, and it was infrared spectroscopy that proved that each of these regions were in fact different in type. The colourless area proved to be type IIa, the yellow outer area type Ib and the blue areas within the crystal were type IIb (with some intermixed type Ib).[38]

The phenomenon of multiple type diamonds is not unusual. Mixed types in diamond have been recorded previously in De Beers synthetic diamonds. While the majority of the yellow De Beers synthetic diamonds tested proved to be type Ib as expected, more than mere confirmation was achieved by these tests.

Natural type Ib diamonds often contain small traces of type Ia content. This was absent from the De Beers synthetic diamonds, consequently aiding in the verification of synthesis.[54] The infrared spectroscopy also identified the cause of the colour variations of the sample, being brownish-yellow, yellow and greenish-yellow as previously mentioned. The brownish-yellow synthetic diamonds were found to have the highest level of nitrogen in the lattice, the yellow synthetic diamonds contained less nitrogen in their lattice and the greenish-yellow synthetic diamonds were found to contain the least quantities of nitrogen.[54]

The observation of the mixed type synthetic De Beers diamond occurred following a closer look at the greenish-yellow stones. Upon magnification, scientists observed unusual colour zoning consisting of growth zones appearing yellow, colourless and distinctly green respectively. The infrared spectrum of the green sections revealed weak bands that suggested a type IIb lattice, with boron impurities present. It was speculated at the time that the boron and nitrogen atoms were dispersed throughout these green sections in a kind of mosaic pattern.[54]

The infrared spectrum of the Russian yellow synthetic diamonds indicated a similar phenomenon. As-grown specimens consisted of a mixture of variable proportions of type Ib and type IaA (nitrogen dispersed as pairs through the lattice). Additionally some of these samples proved to contain small traces of type IaB.[33] This is most unusual, as most nitrogen-containing synthetic diamonds are type Ib. This was mir-

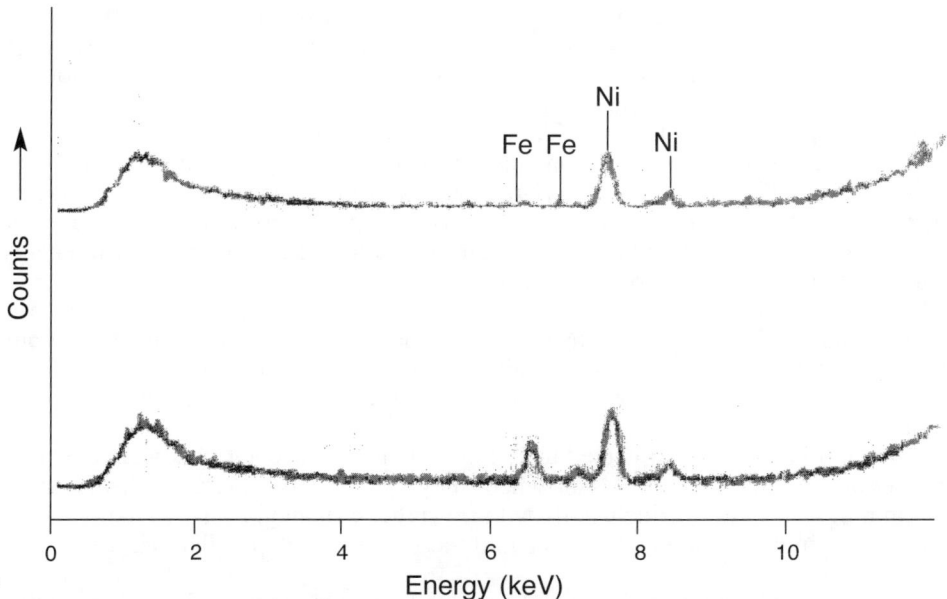

Figure 9.2 EDXRF spectra showing presence of nickel and iron in the metallic inclusions of two synthetic diamonds. (Courtesy *Gems and Gemology* **29**, 4 © GIA 1998.)

rored in the results recorded by the infrared spectral analysis of another Russian synthetic diamond, as reported by Federico Sosso at the Istituto Gemmologico Italiano (IGI) in 1993.[55]

The treated Russian synthetic diamonds, however, did not give the same infrared spectrum as the untreated samples described above. These specimens appeared to be almost completely type IaA, with only a small proportion of type Ib.[33] The Russian synthetic diamonds were the first gem-quality synthetic diamonds to be examined by the GIA that showed a mixture of type IaA and type Ib.[33]

The manifestation of mixed type in synthetic treated diamonds is not unique to Russian diamonds. Outlined in Chapter 9 is the example of treated red synthetic diamonds submitted to the GIA in New York, that proved (with infrared spectroscopy) to be of mixed type. These treated red synthetic diamonds were characterized by the GIA as Ib + IaA + IaB in the case of the 0.55 ct round brilliant cut sample (with IaA >> IaB, and Ia > Ib); and Ib + IaA + IaB in the case of the 0.43 ct radiant cut specimen (with IaA >> IaB, and Ib > Ia), respectively.[61] This is a marked contrast to the classification of natural pink to red diamonds which are either type Ia or IIa.[61]

Natural diamonds, while usually of consistent type have also been recorded as having mixed type, though no natural diamond has been found to consist of types Ib, IIa and IIb together in the one specimen.[38]

EDXRF

This is another specialized method employed to decipher the chemical code of materials under test. In many ways it provides similar and complementary information to that obtained by infrared spectroscopy, with an additional twist. The principle is similar in many ways to the one mentioned, with both relying upon electromagnetic radiation beyond the visible range.

EDXRF, or energy dispersive X-ray fluorescence, is used to determine the chemical constituents of materials by using the X-ray fluorescence and location of absorption edges to identify the elements present in a sample, and prescribe quantities of these elements. This is possible because the innermost core-electron energy levels are not strongly perturbed by the chemical environment of the atom.

Rather, the electric fields acting on these electrons are completely dominated by the nuclear charge. This means that the X-ray spectra of these electrons will be the same regardless of the atom's environment. These energy levels are nearly the same whether the atom is in a complex molecule or a dilute gas. Naturally these energy levels are unique to each particular element.

The quantity of a particular element is determined by measuring the difference in the X-ray absorption, in the regions just above and below the absorption edge for any given element. Even more detailed results can be achieved if optics are used to focus the X-rays onto a small spot on the sample. By doing so the spatial location of a particular element can be obtained.[62-65]

So by applying this complicated technique scientists can determine what elements are present, in what quantities, and where they are located. This is an ideal way of determining the nature of inclusions within synthetic diamonds. This type of test is now routine in gemmological research and the field of synthetic diamond analysis is no exception.[53]

Information relating to the chemical composition of the metallic inclusions within synthetic diamonds gives clues as to the catalyst used and perhaps answers as to unexplained spectral phenomenon. For example De Beers synthetic yellow diamonds were found to contain inclusions composing 60 weight per cent iron and 40 weight per cent nickel. This is also a valuable tool in identifying synthetic diamonds and separation from natural diamonds as no metallic inclusions of this type have been found in natural diamonds.[54]

The visual appearance of such inclusions and other matters relating to the identification of synthetic diamonds is the subject of the next chapter.

Chapter 10

Identification of synthetic diamonds

The prospect of synthetic diamonds infiltrating the jewellery market has become a reality. While the high-purity, near-colourless specimens produced by the major synthetic diamond producers General Electric, De Beers and Sumitomo still elude the grasp of the diamond buying public, Russian near-colourless synthetic diamonds have made a grand entrance,[66] albeit a slow one![67]

One popular test used by jewellers all over the world to assist in the verification of diamond is the use of the diamond probe, or diamond tester. This easy to use instrument is based on the thermal conductivity of diamond, a property not present in diamond simulants, until recently. Mentioned in Chapter 1, moissanite (and its synthetic counterpart, synthetic moissanite) is the exception to this rule.[68] This material is made of silicon carbide and therefore conducts heat (the carbide component being conductive). Both diamond and moissanite (when extremely pure) share a similar thermal conductivity value of around 2300 W/m°C.[56]

The potential problem of synthetic diamonds entering the gem market undetected is now facing the gem diamond industry. Natural diamonds and synthetic diamonds both conduct heat and therefore register on the probe as diamond. Such instruments do not have the capability to differentiate between natural and synthetic diamond, and are therefore useless for that purpose. Additionally the universally acknowledged 'see-through' test will identify cubic zirconia, and differentiate it from a diamond, but will not identify a synthetic diamond. This is because, as expected, the critical angle (~24.43 degrees) is the same as natural diamond and therefore the cut will eliminate the see-through effect in both cases, synthetic and natural.

Where does the average gemmologist turn? The visual appearance seems identical, but is it? The adamantine lustre, extreme hardness, scintillating brilliance and dispersion are all consistent with natural diamonds, and will not give any clues. The first indication of distinction comes when the commonest procedure is applied to diamonds in the jewellery industry (following the application of the well-known diamond probe). That is the grading of the stone, especially the grading of the diamond's clarity.

When undertaking the grading of diamonds, the pressures required to obtain quick and efficient results do not always allow for the detailed investigation of the exact nature of the inclusions themselves, especially when a whole parcel of diamonds is involved. Therefore a closer examination may have to be adopted as standard procedure in the future. We have briefly mentioned so far the presence of inclusions in synthetic diamonds, and will now examine this area in more detail.

Inclusion is by definition, 'the state of being included'. In relation to gemstones, and in particular diamonds, an inclusion is defined as 'particles of foreign matter, solid liquid or gaseous, enclosed within a gemstone'.[51] Our knowledge of inclusions in natural diamonds is as impressive as it is extensive. Science has categorized many types of inclusions in natural diamonds, consisting of:

- protogenetic inclusions such as distorted crystals of diamond, or of other minerals such as olivine (peridot) within the host diamond (formed prior to the host crystal);
- syngenetic inclusions such as well-developed crystals of diamond and graphite (formed at the same time as the host crystal);
- epigenetic inclusions such as alterations to the protogenetic and syngenetic inclusions, cleavages and stress fractures (formed subsequent to the formation of the host crystal).

Almost all of the inclusions in natural diamonds are closely related to the initial conditions of formation. A more detailed description of inclusions in natural diamonds can be found in the references contained within the notes, for those that are interested in this area.[69]

Synthetic diamonds therefore form a class of their own. The fact that they are formed in such a foreign environment, so far removed from the conditions under which natural diamonds are able to form, suggests that the possible manifestation of the inclusions mentioned above within synthetic diamonds is almost negligible. To understand the inclusions that will be found within synthetic diamond we must look at the conditions of formation. These involve a specialized high-pressure, high-temperature device, in the presence of a metallic catalyst. It is from this fact that all the information required to effectively identify synthetic diamonds can be acquired.

Synthetic inclusions in synthetic diamonds

From the definition of inclusions, and our knowledge of the conditions of growth of synthetic diamonds, the state of the inclusions to be found in synthetic diamonds is predictable. Metallic inclusions are a shadowy reminder of the metallic catalyst within the synthetic diamonds crystallized. This remnant of the catalytic solvents suggested to De Beers during their early development which metals were appropriate to stimulate graphite to diamond conversion.[2] Up to 0.2 per cent of the total stone was found to be nickel in the early General Electric synthetic diamond grit.[51] Infrared spectroscopy and EXDRF analysis are now considered common place and therefore the constituents of the metallic inclusion within synthetic diamonds are widely known and published.

What do these metallic inclusions look like? When viewed with a binocular microscope they appear opaque and dark in transmitted light and shiny and reflective in incident light.[46] The actual shape and appearance of such inclusions may vary between crystals and manufacturers. They may appear as large metallic particles, regularly shaped reflective pieces such as platelets,[5] pinpoint inclusions or clouds of tiny metallic residue[51] and vein-like structures.[36] Additionally researchers have observed with standard gemmological magnification, fine rod-like inclusions and unusual triangular inclusions.[53] Individual reports vary in description and conditions under which the inclusions are best viewed.

Collectively there are some generalizations that can be drawn as to the overall appearance of metallic inclusions in synthetic diamonds. The GIA have developed a description of synthetic diamond inclusions based on their extensive experience of synthetic diamonds under magnification. Most synthetic diamonds are found to contain elongated metallic solids with rounded edges, usually in groups.[46] Such inclusions occur in both near-colourless and yellow synthetic diamonds from Russia[33] and De Beers, and may look like a crystal due to the shape of the cavity in which they are situated, not the habit of the metal itself.[54] They are, however, found by the GIA to be more numerous in near-colourless Russian synthetic diamonds, than near-colourless synthetic diamonds from other manufacturers.[46]

Dendritic type inclusions have also been observed in near-colourless synthetic diamonds from Sumitomo, accompanying the usual elongated inclusions described previously.[46] Interestingly this is not the only manifestation of dendritic style patterns in synthetic diamonds. Dendritic surface markings have also been reported on the octahedral crystal faces of De Beers synthetic diamonds.[54]

The second most often reported inclusion in synthetic diamond are 'pinpoint' inclusions, which appear as white dust-like phenomenon. These tiny specks of the metallic catalyst are usually very numerous. They were observed in one third of the 51 near-colourless synthetic diamonds analysed by the GIA (reported in 1997).[46] They have been reported in Russian,[33] De Beers[54] and Sumitomo synthetic diamonds.[36] The main difference was that the pinpoints in the De Beers stones were confined to the growth sector boundaries[54] but those within the Sumitomo stones were randomly distributed.[36] Clouds and haziness seen in synthetic diamonds are a fine distribution of these inclusions.

In De Beers synthetic diamonds the alignment of the inclusions to the internal growth sectors of the crystal has been observed. It occurs with the elongated solid inclusions and pinpoints in both yellow[54] and near-colourless to blue De Beers synthetic diamonds.[70] De Beers synthetic diamonds have also shown other unusual characteristics relating to their inclusions. Eight of the near-colourless synthetic diamonds tested by the GIA, from De Beers, were found to contain thin translucent triangular inclusions. These were found oriented in groups or clusters, visible only in certain directions. Indeed only with advanced analysis could the team secure a description of these inclusions. Although they may be related to weak internal strain, their exact identity is still not understood.[46]

Triangular inclusions have been observed in General Electric isotopically pure diamonds. These were more tabular in nature, and could easily be seen in relief.[53] Similar triangular inclusions, along with rod-like particles of the metallic catalyst, were observed in near-colourless and greyish-blue General Electric synthetic diamonds. No prominent inclusions were apparent in the yellow samples from the same source.[37] These data are significant in the identification of synthetic diamonds and in the separation from natural diamonds, because natural diamonds do not have metallic inclusions of these kinds.

The visual appearance of metallic inclusions seen in synthetic diamond inclusions is not like the crystallites of natural diamonds.[36] These when transparent, will serve to identify the diamond as a natural stone. Opaque inclusions in diamond should be tested further, and checked for the reflective metallic lustre.[52] While metallic inclusions have been reported in natural diamonds,[71] this very rare occurrence does not have the appearance of the type outlined here.[53] The presence of reflective lozenge shaped metallic inclusions is a strong indication that the stone may be a synthetic, and further testing should be undertaken.[52]

Regardless of the shape, size, distribution or physical appearance of the metallic inclusions in synthetic diamonds, there is one feature that is common to all. This is magnetism.

Magnetic attraction

Being metallic, the inclusions within synthetic diamond are not merely grey and shiny, as expected of many metals, they are also magnetic. Magnetism is a phenomenon associated with the motion of charged particles such as electrons. It involves magnetic fields, which are regions where a force is exerted on a magnetic body, and the effects of such fields.

Magnetism has been recognized since ancient times. The mineral magnetite (lodestone) is an oxide of iron that has the property of attracting iron objects, and has been known throughout the ages, dating back to the Greeks, Romans and Chinese. The most familiar evidence of magnetism is the attractive or repulsive force observed to act between pieces of magnetic materials such as iron. When iron is stroked with magnetite, it acquires the same ability to attract other pieces of iron. The resulting magnet thus produced is polarized; that is, it has two extremities described as north-seeking and south-seeking poles.

The strong magnetic property of iron, cobalt or nickel, is known as ferromagnetism. These three prominent ferromagnetic metals are 10 000 times more potent than copper. The importance here is that iron, nickel and cobalt are among the most favoured metals used as catalysts in diamond synthesis, and therefore constituent inclusions within synthetic diamonds.[52] Why are these metals attracted to magnets?

In certain types of solids, the atoms possess a permanent magnetic moment, or attraction (they act like tiny bar magnets). In most solids, the directions of these moments are random. In ferromagnetic solids (such as those listed above) the moments of neighbouring atoms spontaneously align in the same direction when placed in the vicinity of a magnet. This lining up of all of the moments (in this case of an inclusion in a synthetic diamond) causes the metal to become magnetized, with exceptionally high energy. This occurs when a ferromagnetic metal inclusion in a synthetic diamond is magnetized. The 'moments' in the metal inclusions are aligned, forming a tiny magnet that is subsequently attracted to other such magnets.[56]

This property is of great use in the separation of natural and synthetic diamonds.[37] Whether the synthetic diamond has large reflective elongated metallic inclusions, pin points of catalyst, fine wispy clouds of microscopic pinpoint inclusions or no visible inclusions at all, the presence of residual metallic catalyst (in any of these forms) will be sufficient to cause the diamond to be attracted to a strong magnet. This type of attraction can be seen if the diamond is suspended by a fine thread, away from air currents, and a magnet brought close to (but not touching) the stone. If movement of the stone (due to magnetic attraction) is observed, the diamond is then known to be synthetic.[52] Be careful how you mount the stone for testing however – wire and Blu-tack® are also magnetic![73] Details on suggested methods are contained in the next chapter.

The magnetic element iron has been located in natural diamonds, within natural crystal inclusions of minerals such as garnet – and diamond polished on an iron polishing wheel or scaife may be contaminated with tiny amounts of iron.[52] These amounts however are insignificant in comparison to those distributed within synthetic diamonds, and are insufficient to instigate attraction of energy high enough to register during the above test.

Graining

Before moving on there is another characteristic of synthetic diamond microscopy that should be discussed. One in which like manifestations of some of the forgoing inclusions are also related to angular formations and internal growth sectors – graining.

Graining is the materialization of the crystal's growth, with the graining marks representing growth sectors, whether in natural diamonds or synthetic diamonds. They may be irregular, linear or planar and may be internal or external. External graining is influenced by internal graining, despite changes made to the stone such as polishing or faceting. Faceting marks are a result of different circumstances (in this case applied solely by man). Current thought is that internal graining is a direct result of lattice orientations, or structural irregularities, which may be influenced by the presence of inclusions.[73] It has also been suggested that graining is due to slight fluctuations in the refractive index of the diamond in neighbouring parts of the lattice. This change in the refractive index is responsible for the optical contrast that we see as graining.[54] Graining in natural diamonds has been well catalogued. It is evident in a number of recognizable forms.[69] We will discuss these briefly in comparison with the graining found in synthetic diamonds.

Graining in synthetic diamonds seems to be confined to coloured synthetic diamonds, with none of the near-colourless synthetic examples tested by the GIA displaying graining. In these samples, only a few showed any form of colour zoning under strong magnification.[46] This was the case in separate studies of the isotopically pure synthetic diamonds from General Electric,[53] and the colourless-to-blue De Beers experimental synthetic diamonds, which showed graining only very slightly on the intersections of some of the coloured areas.[70]

Yellow synthetic diamonds, however, are entirely different in this regard. Graining in these stones is distinct, both internally and externally. As discussed, the morphology of synthetic diamonds is unique, combining the cubic and octahedral forms, and, more importantly, changing between the two during growth. The internal graining in synthetic diamond is another feature exhibiting this unusual growth to the observer.

Sumitomo, De Beers and Russian yellow synthetic diamonds have all exhibited internal graining, including the treated Russian specimens. Interestingly the same phenomenon has not been observed in General Electric yellow synthetic diamonds.[36]

There are three main types of internal graining characterized for synthetic diamonds:

- hourglass graining
- radiating or parallel graining lines
- stop-sign graining.

Hourglass graining manifests as a result of graining along the intersections of the cubic and octahedral growth sectors. It is distinctive and is best viewed through the pavilion of a faceted stone.[52] This type of graining has been observed in De Beers samples[54] and Sumitomo faceted synthetic diamonds. Sumitomo synthetic diamonds displayed a slightly different type of graining.[36]

Unfaceted Sumitomo yellow synthetic diamonds were found to have graining appearing as lines parallel to the outer shape of the crystal and radiating lines emanating from the centre of the stone outward. These lines together formed V-shaped areas like an 'iron cross'.[36] Similar lines were reported in Russian synthetic diamonds, with rectangular graining lines radiating outward, in addition to intersecting grain lines.[33] Lines in De Beers yellow synthetic diamonds were found to exist in parallel to the cubic faces, and were well developed in cubic sectors.[54]

The third type, the stop-sign shaped graining, was also evident in synthetic diamonds from more than one manufacturer. Sumitomo synthetic diamonds exhibited the same octagonal stop-sign graining internally and on the surface,[36] as seen in the De Beers samples,[54] and the Russian synthetic diamonds.[33] This type of graining does not occur in natural diamonds, and therefore may be considered diagnostic for synthetic diamonds.[54] Interestingly, both graining and growth zoning was observed in the early General Electric samples supplied to the GIA.[37]

In fact, graining of the types described are all considered diagnostic for synthetic diamonds. They are not seen in natural diamonds, which do not show graining patterns such as these. Coloured graining, or colourless phantom graining parallel only to the octahedral directions, indicate that the stone is a natural diamond.[52]

It has been suggested that the graining in yellow synthetic diamonds may be attributed to the presence of nitrogen, explaining why it is evident in yellow type Ib synthetic diamonds but not in nitrogen-reduced samples.[70]

Grading synthetic diamonds

This discussion inevitably leads to the question of the grading of synthetic diamonds. What level of purity is observed? What clarity grades can be expected? Just how white are the near-colourless synthetic diamonds available on the market today?

The gem diamond industry is most likely to be confronted with near-colourless synthetic diamonds, as colourless diamonds are more popular in the jewellery market place. At the time of writing, colourless synthetic diamonds exceeding 1.00 ct in size have not been faceted. Near-colourless or colourless synthetic diamonds are of type IIa or mixed type IIa + Ib + IIb. The colour, while appearing colourless, may be very light grey, very light blue, light yellow or very light green. This colour may appear in sectors, and is best identified when colour grading the stone, as the very light colour may be evident against the colour of the master stones.[37]

Initially it should be said that the stance taken by the Gemological Institute of America Gem Trade Laboratory is one of non-participation. To this end the GIA GTL does not provide clarity and colour assessments for synthetic diamonds to the public. Any colour and clarity grading performed during the course of research has been solely for the purposes of compiling a frame of reference to describe said diamonds. Those grading activities performed by the GIA to date have been completed using standard grading procedures and instrumentation.[46]

Of the 51 near-colourless synthetic diamonds mentioned previously, only 26 were colour graded. Of these 11.5 per cent fell within the range of F to G colour on the GIA scale. Forty-six per cent were ascribed the colour grade of H, with the group forming the majority of the tested stones. A further 31 per cent of those tested were graded as being I in colour, and the remaining 11.5 per cent were found to be J to K in colour. The term 'near-colourless' has been applied to all of these stones.[46]

Clarity of these synthetic diamonds was not tested, nor is it described within the original publication. However investigations undertaken by the author in the acquisition of synthetic diamond samples revealed that synthetic diamonds were advertised as being available in clarity ranging from VVS to VS and culminating in Piqué. For example, three near-colourless to blue De Beers synthetic diamonds examined by the GIA were graded as VS_1, SI_1, and SI_2.[37] The majority of the colourless synthetic diamonds obtained by the author were in the SI_2 to P_1 range, with graining, inclusions and magnetism as described here.

Chapter 11

Gemmological testing

As we have seen in Chapter 10, there are a number of ways to identify synthetic diamonds during the course of a standard grading exercise. However, what if the diamond does not display the reflective inclusions, or it is a near-colourless synthetic diamond without the hour-glass and stop-sign graining? What if you do not have a magnet handy? In the course of a laboratory report, a grading exercise or a testing procedure, gemmologists may in the future be confronted with a stone that they suspect is a synthetic diamond, even if this is not supported by the preliminary microscopic examination. Where to turn?

Thankfully advanced studies conducted on many synthetic diamonds over the last 20 years have determined that there are a number of diagnostic features of synthetic diamonds that can be easily identified with standard gemmological instruments in the back office or small scale laboratory. As these tests are based on the same gemmological characteristics outlined in Chapter 8, the results will be familiar to the average gemmologist and well within the understanding of all. However these tests are for the most part rarely performed on diamonds. The use of the diamond probe and the visual properties are usually sufficient to identify the stone as a diamond, but not to identify it as a synthetic diamond.

For the first such test we will return to the microscope to observe another phenomenon, one that may have escaped our attention during the preliminary grading evaluation, and should therefore be included in all grading activities in the future. This is the search for the elusive colour zoning.

Colour zoning

We know that synthetic diamonds are grown as yellow type Ib stones, near-colourless type IIa stones, blue type IIb stone and stones of mixed type with slightly variable colour. What has now become clear is that the colour is not evenly distributed within the stone. This is probably best seen in the yellow synthetic diamonds, as the contrast in which the coloured zones is more dramatic – although observation may depend upon fashioning, the crystal itself and the growth conditions.[38]

Colour zoning is another manifestation of the internal growth structure of the synthetic diamond crystal. This is in many ways a more visual and easily 'read' revelation. Additionally colour zoning has been found to exist in close relation to the previously mentioned graining, and they may be observed together. Just as colour is caused by the presence of impurities (Chapter 7), so colour zoning is caused by irregular distribution of these impurities during growth.[54]

For example, in the case of the yellow synthetic diamonds from Sumitomo, uneven distribution of colour was observed as early as 1986. Unfaceted (or rough) pieces of synthetic diamond exhibited a deep yellow inner zone and a narrow near-colourless outer zone, although overall the stone appeared a rich yellow colour. In addition to this, more subtle intensity variations were discovered. This colour zoning was curiously found to be substantially less obvious, or not visible at all in the faceted stones analysed at the same time.[36]

Similar colour zoning was observed in Sumitomo synthetic diamonds tested in 1992. Here variations in colour zoning were characterized in terms of the crystallographic orientation of said zones. It was found that the fine dodecahedral sectors {110} were blue in colour, while the more prominent octahedral sectors {111}, closer to the edges, were yellow. Even more interestingly the inner sector of the stone was found, by the use of infrared spectroscopy, to contain less nitrogen than the other sectors, and was colourless. The synthetic diamonds with these zones appeared a yellowish colour overall.[38]

De Beers yellow synthetic diamonds were also found to have distinct colour zoning, oriented with respect to the internal growth structure as well as the outer shape of the as-grown crystal. The colour zones in these stones were described as geometric with a regular pattern. Near-colourless veins extending inwards from the edges of crystals of the De Beers material were located beneath cubic (100) or dodecahedral (110) faces.[54] It was determined at the time, that these colour zones were probably caused by differential distribution of the nitrogen within the crystal.[54]

One fact is clear, however, when all of the data and observations from specimens of the known synthetic diamond manufacturers is complied. Whether the colour of the original crystal is defined as brownish-yellow, greenish-yellow or yellow overall, the zoning is always distinct.[54]

Colour zoning in the Russian yellow synthetic diamonds was equally distinct. The colour zoning in these samples was described as near-colourless to light yellow, with a square or rectangular central area surrounded by darker yellow sections oriented in a four-fold symmetrical arrangement. Interestingly the colour zoning was found to be more prominent in the treated Russian synthetic diamonds (the treatment of which is outlined in Chapter 13).[33] Similar colour zoning has also been observed independently in yellow Russian synthetic diamonds in other laboratories around the world.[55]

The other type of treated synthetic diamonds outlined in Chapter 13, the treated red coloured synthetic diamonds, also displayed distinctive colour zoning. Through the crown of one sample a square and cross-shaped light yellow zone was evident, surrounded by a larger red area. Here the yellow areas were found to be tabular and narrow compared with the red areas.[61] The other sample also exhibited similar yellow and red coloured zoning.[61]

Colour zoning has been described by the GIA Gem Trade Laboratory as the most prominent characteristic of yellow synthetic diamonds.[54] But what of near-colourless synthetic diamonds? How does one see colour zoning when the stones lack colour?

The experimental blue-to-colourless synthetic diamonds from De Beers (doped with boron to produce the colour) certainly have the diagnostic colour zoning we have come to expect from synthetic diamonds. In diffused transmitted light it was determined that definite colour zones existed consisting of light blue, darker blue, light yellow and near-colourless respectively. The blue zones were found to be the most distinctive, appearing as parallelogram (wedge) shaped areas with sharp edges and acutely angular corners. These blue zones were oriented beneath the cubic (111) faces of the original crystal. The shapes and outlines of the remaining yellow and near-colourless zones was less apparent, but none the less present.[70]

Gemmological testing 99

No colour zoning has been observed in the isotopically pure carbon 12 variety of synthetic diamond produced by General Electric.[53] Of the 51 near-colourless synthetic diamonds tested by the GIA, from various sources, none revealed any colour zoning when examined under normal conditions.[46] Five of the samples (all originating from De Beers) exhibited faint yellow or blue colour zoning under magnification.[46]

Hence from this we can deduce that while colour zoning may offer a definitive tool in the identification of coloured synthetic diamonds, natural diamonds show colour zoning in a planar or roiled effect and not, as synthetic diamonds do, aligned to the internal growth sectors.[52] In the case of near-colourless synthetic diamonds, however, the presence and visibility of colour zoning cannot be relied upon to denote the existence of cubo-octahedral growth sectors unique to synthetic diamonds. There are other tests that help determine the existence of such internal sectors, and lead to the identification of synthetic diamonds and their separation from natural diamonds.

Luminescent patterns

The observation of the ultraviolet fluorescence of diamonds is regularly performed by diamond graders, as part of the grading exercise. Long wave ultraviolet light sources are included as part of many diamond colour grading light sources and are therefore part of most gemmological laboratories. Diamond grading reports and certificates include a description of the fluorescence of the diamond under test, detailing the intensity of the colour seen. To this end the diamond is best viewed in partial darkness and from the side of the pavilion. Until now this technique has supplied all of the information required on a daily basis for those in the diamond and jewellery industries.

For the purposes of identifying a synthetic diamond, more information is required. As we have seen in Chapter 8, the fluorescence of synthetic diamonds is unique and important in the description of the stone. The colour and intensity of the fluorescence is important, just as it is in the case of diamond grading, but the patterns also become important. In addition to this, the standard use of the long wave (365 nm) ultraviolet light source is no longer sufficient; one must also employ the short wave lamp.

Short wave (253.7 nm) ultraviolet light is of importance because most synthetic

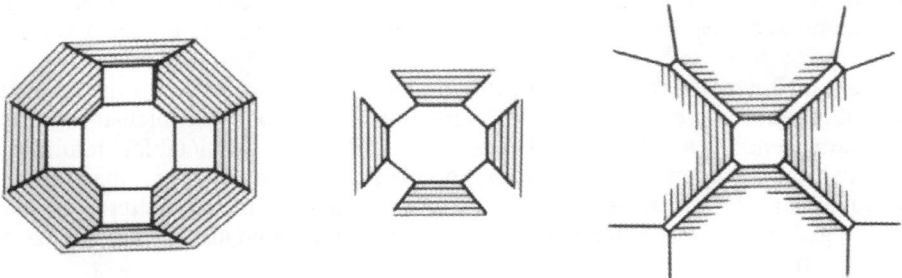

Figure 11.1 Three configurations typical of colour zoning and luminescence patterns seen in synthetic diamonds. Partial and combinations of these patterns are also observed. (Courtesy *Gems and Gemology* **29**, 4 © GIA 1998.)

diamonds fluoresce more strongly under short wave ultraviolet light than they do under long wave ultraviolet light.[52] Others do not show a reaction to long wave ultraviolet stimulation at all.[38] We know the fluorescence of synthetic diamonds is uneven, just as we have found that the colour is uneven. Instead patterns are evident, or a general zoning – all of which constitute signposts in the separation of natural diamonds and synthetics.

Fluorescence, just as colour zoning, is evidence of the internal growth sectors within the original crystal. Therefore to enable to use the patterning in identification we must think of it in terms of these sectors.[46] As we have discussed these patterns may be cross-shaped,[46] octagonal,[52] or in the form of planar banding.[41] Can you recognize these shapes? Indeed they are very similar to the hour-glass, stop-sign and parallel graining described in Chapter 10. Again they follow the intersecting cubic and octahedral growth sectors, highlighting the differential growth of the crystal within the high-pressure device.

The distribution of the patterning may vary. Remember the Russian synthetic diamonds in which the fluorescence to ultraviolet light was observed only in certain growth sectors.[36] The colour also may vary. Sumitomo yellow synthetic gem-quality diamonds exhibited weak orangy-yellow fluorescence, and yellow fluorescence in two stones tested at the same time.[38]

For the purposes of separation it is not imperative to memorize all of the patterns and colours possible in synthetic diamonds. As new techniques emerge, and properties vary between manufacturers, keeping up with the combinations alone could become a full-time occupation. The use of short wave ultraviolet light, and the strength of the fluorescence under this type of UV light, when compared to the fluorescence under long wave UV is of importance, and attention should be given to this test. Here the comparison of results is the key to aid in separation, rather than the level of intensity (to either long wave UV or short wave UV).[52]

The ultraviolet fluorescence of natural diamond is typically more pronounced under long wave UV than short wave UV. The difference in intensity under the two wavelengths is obvious, with an ample contrast easily visible. A diamond displaying a very strong long wave UV reaction, and a weak short wave UV reaction will probably be a natural diamond (although further verifying tests are recommended). The majority of synthetic diamonds fluoresce more strongly under short wave UV than long wave UV. This is always the case with near-colourless synthetic diamonds.[52] In the cases where the visible reactions are of similar intensity, or the long wave UV reaction is only mildly more pronounced than the short wave UV reaction, further tests are still necessary. This occurs in the rare cases of yellow coloured synthetic diamonds. The diagnostic cubo-octahedral patterning should be examined next.[52]

Overall, however, the pattern is far more important than the colour, or the intensity. Natural diamonds can exhibit a rather diverse range in colour and intensity in ultraviolet fluorescence, but the pattern is usually consistent and predictable. Natural diamonds exhibit an even (usually blue) fluorescence, although there may be fine concentric bands.[41] Synthetic diamonds show unusual geometric patterns (usually yellow to green) in a cross shape, octagonal shape or a combination of glowing zones and inert zones.[52]

The results for natural diamonds and synthetic diamonds are very different. The smooth growth of the natural diamonds produces smooth reactions to ultraviolet radiation, while the violent and hastened growth of synthetic diamonds produces striking geometric patterns. This is a conclusive test, and evidence of these patterns is diagnostic.[52] Sometimes, however, it is not the dramatic results of this technique on synthetic diamonds that assists in separation, but the obvious lack of a result.

Separation with the synthetic spectrum

Chapter 8 discussed the spectral features associated with synthetic diamond. There we learned that the GIA observed no significant spectra in the majority of the 51 near-colourless synthetic diamonds referred to throughout this text.[46] Remember, however, that in diamond types assigned to synthetic diamond, type Ib, IIa and occasionally IIb, the lack of spectrum is expected.[52] This fact alone can assist us in identifying synthetic diamond, based on the rarity of these types in nature, especially if the diamond under test is near-colourless. Near-colourless synthetic diamonds which do not exhibit any spectral features can provide a hint to their identity, because natural near-colourless diamonds do exhibit diagnostic absorption bands.[51]

To assist in the explanation of spectral features in this section we have utilized transmittance visible range spectrophotometer data from the Adamas Gemological Laboratory in the United States. The instrument used to produce these figures was the SAS2000 Spectrophotometer Analysis System for diamond and gemstone evaluation and grading. This will be explained in more detail in Chapter 12, along with the introduction of some of the comparative results of the system. Shown below, however, in Figures 11.2 and 11.3, are examples of how the system is used to identify spectral features, along with their relative intensity. Figure 11.2 depicts the transmittance data for a series of natural colour grading master stones. Note that the increasingly larger 'troughs' or valleys at the left side of the figure correspond to the absorption bands we would seek with a hand-held gemmological spectroscope. Figure 11.3 compares the

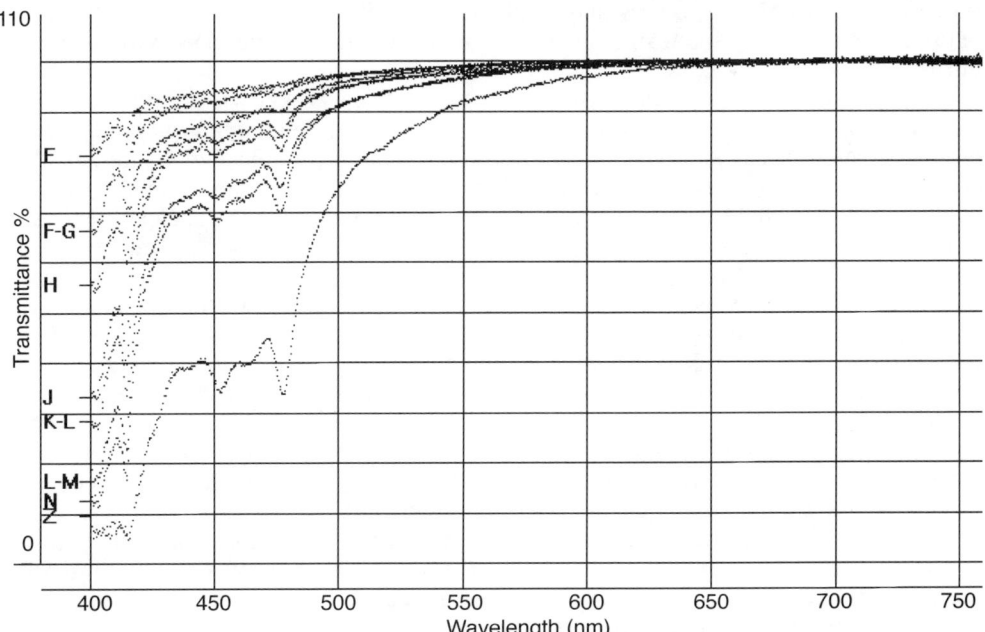

Figure 11.2 SAS2000 spectrophotometer transmittance data for a range of natural diamond colour master stones. (Courtesy Adamas Gemological Laboratory © 1998.)

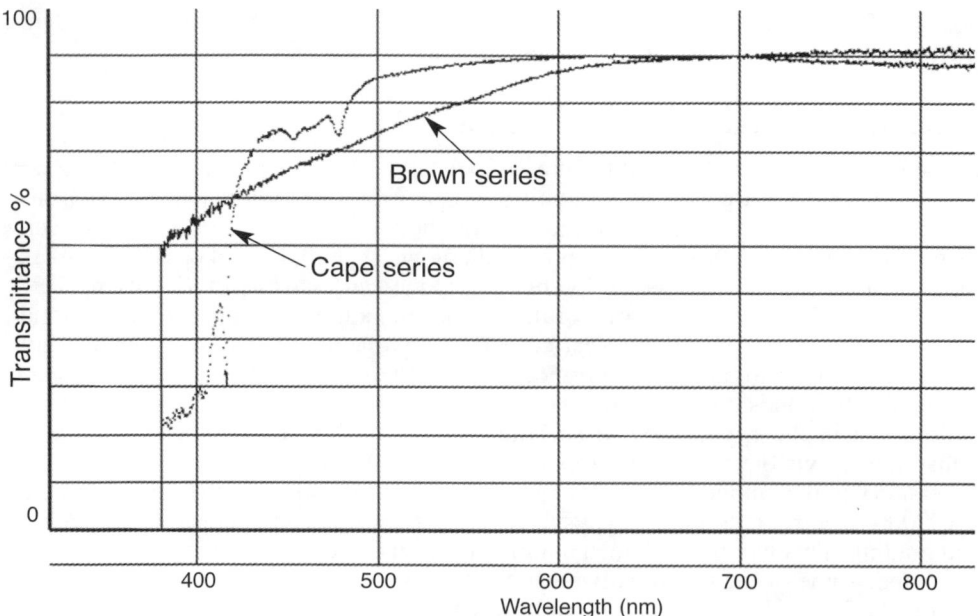

Figure 11.3 SAS2000 transmittance spectra for a typical cape series natural diamond (curved spectra) and brown series natural diamond (more linear shaped spectra). (Courtesy Adamas Gemological Laboratory © 1998.)

transmittance for Brown Series and Cape Series natural diamonds. Although these natural diamonds may seem similar to the naked eye, one can see the obvious differences in the spectra here.

In 1891, B. Walter was the first person to observe the well known cape spectrum in natural diamonds. This series of bands associated with type Ia natural diamonds (that are usually colourless and have a blue fluorescence to ultraviolet light), are usually visible in the blue to violet part of the spectrum. The most prominent line is that at 415.5 nm, deep in the violet. Other important lines associated with the cape spectrum are found at 478.5 nm, being the second most conspicuous line, and at 465 nm, 435 nm and 423 nm.[5] To improve visibility of these lines, a blue filter, such as a flask containing copper sulphate should be used.[51] The observation of the cape spectrum is confirmation of a natural diamond. The positions of these bands can be discerned in Figure 11.3, with absorption lines manifesting as troughs in the curved gradient of the spectrophotometer transmittance. No synthetic diamonds will show a cape spectrum.[52]

The second group of natural diamonds exhibiting bands in their spectrum are the Brown series of natural diamonds, or those that have been treated to alter their colour. These bands, due to deformation of the diamond lattice, are located at 496 nm, 503 nm and 595 nm.[51] The transmittance of this group of natural diamonds is also shown in Figure 11.3. Note the almost linear progression of increased transmittance towards the red (700 nm) end. Again, these bands are not seen in synthetic diamonds.[52]

Yellow type Ib and near-colourless synthetic diamonds show no spectrum at all, but neither do all canary yellow type Ib natural diamonds. In this case it is imperative to take a closer look at the spectrum. Thankfully, a number of spectral bands have been seen in synthetic diamond. These lines, which appear between 470–700 nm, resulting from the included nickel catalyst (outlined in more detail in Chapter 8), are listed in Table 8.1. Synthetic diamonds have been known to display an increasing general type of absorption toward the violet ends of the spectrum.[51] Remember that the natural canary yellow type Ib diamonds will not show any spectral bands .The presence of these nickel-related features then is an indication that the diamond is synthetic.[52]

In the case of spectral analysis, only the detection of the familiar Cape spectrum in natural diamonds can be treated as definitive and diagnostic. This is because at room temperatures the nickel-related bands in synthetic diamonds can be difficult to see, and human eyes (and brains) have a habit of 'seeing what we want to see', whether present or not. In this case it is wise to verify the result with a further test for clarification. If these nickel bands are seen, then this is an indication of the presence of nickel in the lattice. If this really is the case the stone should react to a magnet as well.

The strain of detection

ADR, or anomalous double refraction, is a feature in some gemstones, both natural and man-made. Strain within the lattice of the material causes a type of distortion, that in turn results in the appearance of a false double refraction in otherwise singly refractive gemstones. Unlike the spectral features described previously, this type of strain exists in both natural and synthetic diamond. Rarely will either natural or synthetic diamonds be devoid of strain. If this does occur, it will not assist in separation.

The ADR visible in natural diamonds appears as dark shadowy bands and mottled patches with alternating bright interference colours, when viewed through crossed polaroids.[51] This can be achieved with either a stand-alone polariscope or polar attachments on the microscope. The diamond is best positioned between tweezers holding the diamond by the table and culet, to enable viewing in the direction of the girdle. This will eliminate residual internal reflections from diminishing the observed effect.

The ADR present in synthetic diamonds, both near-colourless and coloured (yellow and blue), is quiet different. The pattern of strain in synthetic diamond is dull and subdued in comparison with that of natural diamonds. The patterns are cross-like and grey in colour, without the contrast of light and dark bands that are visible in some natural diamonds.[52] This weak strain can be identified by the presence of low order interference figures, usually seen as black, grey and white cross-shaped patterning between crossed polaroids.[46]

Again this is not a conclusive test in its own right, and is best applied in combination with other tests. To enable readers to decipher all of this information, and to provide a sequence or direction for testing, the information has to be tabulated and supplied as in Table 11.1. This is designed to give a systematic approach to the separation of natural and synthetic diamonds. It may eliminate the performance of unnecessary tests, wasteful of valuable time in a retail or wholesale environment. For best results follow the sequence in the chart – the test at the top should be performed first and the tests at the bottom should be performed only if needed. Conclusive tests have been printed in bold text to assist in reading the results.

With the information contained within Table 11.1 it is possible to successfully identify synthetic diamonds and separate them from natural diamonds. Two or three tests in favour of either natural or synthetic stones is all that is required (depending upon

Table 11.1 Tests and suggested sequence recommended for assisting in distinguishing between natural and synthetic diamonds.

Test	Result for natural diamond	Result for synthetic diamond
Microscopy	**Crystalline inclusions**	**Reflective metallic inclusions**
	Octahedral graining	**Hourglass or stop-sign graining**
	Planar graining	**Strong geometric colour zoning**
	No graining visible	
	No inclusions or small identifiable inclusions	
	Even colour distribution	
Magnetism	No reaction	**Movement when exposed to magnet**
		No movement (inclusions too small)
Fluorescence	Blue fluorescence	Yellow–green fluorescence
	Stronger long wave fluorescence	**Stronger fluorescence to short wave then long cross shaped and geometric patterning**
	No fluorescence to short wave ultraviolet	
	Yellow fluorescence	
	Equal reaction to long and short wave ultraviolet	No reaction to long wave ultraviolet
Spectroscopy	**Cape spectrum**	No spectrum
	415.5 nm line	**Lines due to nickel**
	Brown series lines	Increased violet absorption
Polariscope	Banded, mottled strong ADR	Cross like ADR
	Colourful interference figures	Black to grey interference figures

which tests are applied); however, if in doubt a consensus from the whole range of tests may be applicable. Definitive tests such as a metallic inclusion (in the case of synthetic diamonds) or the Cape spectrum (in the case of natural diamonds) may be all that is required.

If you are confronted with a parcel of stones that requires separation, the best approach is to begin with the fluorescence to ultraviolet light and separate those that fluoresce yellow to short wave UV and from those that fluoresce blue to long wave UV, and continue from there.

All of these tests can be successfully applied in normal (or darkened) circumstances, and with instruments that are readily employed by the average gemmologist.[41] However, for those who deal with synthetic diamonds on a regular basis, and have money to invest, there are specialized instruments designed specifically to identify synthetic diamonds, and to isolate them from their natural counterparts. These are the subject of Chapter 12.

Chapter 12

Specialized separation instruments

Responding to the concern within the jewellery trade regarding the commercial sales of synthetic diamonds, researchers at De Beers Diamond Trading Company Research Centre have developed two specialized instruments to enable a simple and reliable separation of natural and synthetic diamonds.[74]

When developing these instruments, researchers kept in mind the needs of the average gemmologist or jeweller. The need was to be able to identify synthetic diamonds with the same ease of use as applying a diamond probe to separate a diamond from a cubic zirconia.[41] Both instruments are based on the location of gemmological features of natural and synthetic diamonds contained within Table 11.1 in the last chapter. These instruments, however, take the guess-work out of identification, by providing the user with straightforward results relating to the stone under test. It must be remembered that current laws and regulations require full disclosure of any enhancement or synthesis to customers at the time of sale.[75, 76] These instruments have been ten years in the making and are named the DiamondSure™ and the DiamondView™. They have been designed to enable the rapid evaluation of a large number of potentially synthetic diamonds, and to assess diamonds already set in jewellery in the case of the DiamondSure™. They are best employed in a two-stage process. It is recommended to use the DiamondSure™ first, followed by the more detailed examination provided by the DiamondView™.[41]

The DiamondSure™

The 2.8 kg DiamondSure™ fits neatly on a desk top and employs the principle of spectral detection in its operation. The diamond(s) under examination is placed in a small black (detachable) stage, on top of a fibre optic probe (diameter 4 mm) with the table of the stone facing down. Stones that are already set in jewellery may be analysed by using the fibre optic probe (with the collar like table removed) like a pen, or standard diamond/simulant testing probe. The instrument seeks the cape spectrum, or more accurately, the 415.5 nm absorption band apparent in the majority of natural diamonds by measuring the intensity of retro-reflected light centred around the wavelengths in the region of this band.

Using the exclusive De Beers software, the intensity data is compared with that corresponding to the result seen in the average natural diamond. The DiamondSure™ is highly sensitive (to 0.03 absorbance units), and testing takes around 4 seconds.[41] Obviously there is more to this than the brief description above. Nothing is as simple as it may seem. To make best use of the DiamondSure™ it is far more important for us to delve into the results obtained by the device, as well as how to read these results

106 The diamond formula

and how to use them as part of a systematic examination of synthetic diamonds. There are far more interested parties in the industry that constitute 'practical operators', rather than developmental scientists who are interested in the inner workings of the sophisticated mechanisms. Those interested in more detail in these areas are directed to the appropriate reference contained within the notes section for this chapter.

The comparison of the established data and the test data determine the presence of the 415.5 nm line in the diamond under examination. If the line is found to be present,

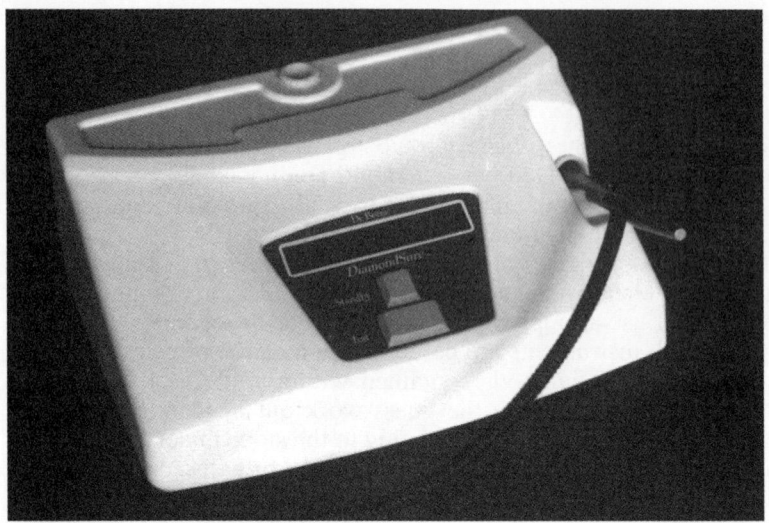

(a)

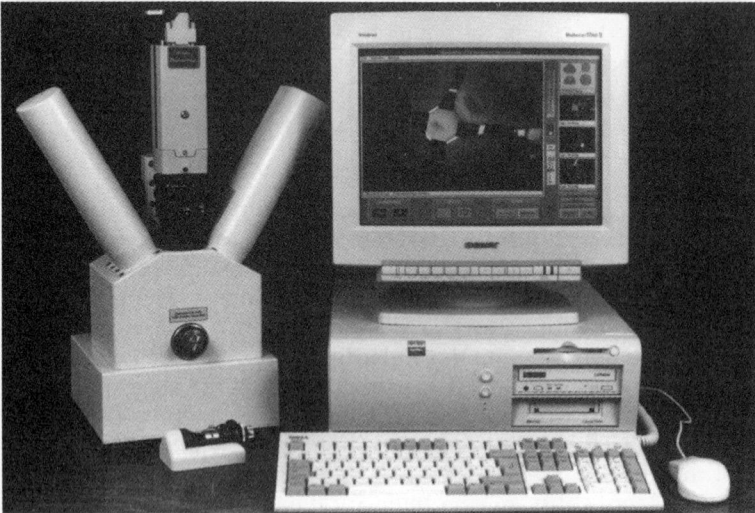

(b)

Figure 12.1 The DiamondSure^{TM} (a) identifies natural and synthetic gem diamonds by detecting the presence or absence of the 415 nm absorption line. If this test is inconclusive, the more expensive DiamondView^{TM} (b), which consists of a fluorescence imaging unit and associated computer, can be used to reveal the diagnostic growth patterns characteristic of natural and synthetic diamonds. (Photographs courtesy of De Beers © 1999.)

the sample stone has been conclusively identified as a natural diamond and no further testing is required. This procedure is undertaken by the operator simply by placing the diamond on the stage (as described above) and pressing the test button on the front of the instrument.

The liquid crystal display above the operation buttons displays three possible messages depending upon the results of the test. These are 'PASS', in which case the diamond has exhibited the 415.5 nm line and is verified as being a natural diamond. This test is conclusive. De Beers researchers detected this spectral feature in 95 per cent of the natural diamonds tested with the DiamondSure™, and in none of the synthetic diamonds tested.

In the case of the minority 5 per cent of natural diamonds that may not show a 415.5 nm absorption line, or a synthetic diamond (which as we know from previous chapters will not show a 415.5 nm line), the message 'REFER FOR FURTHER TESTS' will be displayed.[41] This does not mean that the diamond is synthetic. Note that while the 'presence of a 415.5 nm line' is a conclusive indication for diagnosis, the 'lack of 415.5 nm line' in not! The lack of this line indicates only that a stone *may* be a synthetic, merely that there is a possibility that it is, as some natural diamonds do not exhibit the 415.5 nm line either. For example, type Ib and type II, and some fancy coloured diamonds and high coloured D to E coloured diamonds were found not to show the 415.5 nm line.[41]

It is also important to note here that there are rare cases where synthetic diamonds can be treated to induce the presence of the 415.5 nm absorption line in the spectrum. This treatment involves annealing by exposing nitrogen-containing synthetic diamonds to a process of heating to a temperature of around 2350°C at ambient pressures of 85 000 atmospheres.[77] However, this process is very damaging to the tungsten carbide anvils within the high-pressure apparatus and diamonds suffer during this extreme practice. The surface of the crystals becomes etched and the risk of fracturing is great. For these reasons it is considered unlikely that such treated diamonds will enter the gem and jewellery market place, as the application for such technology for this purpose is not commercially viable.[41]

The third message displayed by the Diamond Sure™ is 'INSUFFICIENT LIGHT'. This is due to the absorption in the region utilized by the instrument being so strong that no light is able to penetrate and reach the detector. This occurs if the diamond is particularly large, or the colour particularly deep. De Beers suggest carefully repositioning the stone over the probe tip, and re-testing. Rest assured, that if the stone is positioned in such a way as to violate the integrity of the test, the DiamondSure™ has a safety mechanism designed to eliminate the chance of a false or incorrect result. In this case it automatically shows the message 'REFERRING FOR FURTHER TESTING'.[41]

During the development of the DiamondSure™, the De Beers DTC researchers tested approximately 18 000 natural polished diamonds, ranging in size from 0.05 ct to 15.06 ct, with the majority of stones between 0.25 ct and 1.00 ct, placing them well in the range of the diamonds most commonly used in the jewellery industry. Colours of the stones surveyed ranged from D to R, as well as some fancy coloured diamonds. Both round brilliant cut and fancy shaped diamonds were tested and 4.3 per cent of stones were 'referred'. In a separate evaluation 60 per cent of D coloured diamonds were referred.[41]

In addition to these natural diamonds, 98 De Beers synthetic diamonds were tested (representing all colours manufactured), together with some Russian BARS-grown synthetic diamonds. Each synthetic diamond was tested 10 times (as a checking procedure) and 100 per cent of the synthetic diamonds were 'referred for further tests'. This extensive testing regime demonstrates the reliability of the DiamondSure™ in normal conditions.[41]

The DiamondSure™ is described by De Beers DTC as a diamond screening instrument, and the power supply is suitable for use in every country. Larger fibre optics can be fitted to accommodate larger stones if required.[41] This makes the DiamondSure™ a versatile and highly valuable piece of equipment for the modern laboratory. But what of the diamonds that register 'refer for further testing'? The researchers at De Beers have developed another user-friendly, compact instrument to assist in the identification of these stones.

The DiamondView™

In the same way that the DiamondSure™ works on the detection of the impurities in the diamond lattice, through the spectral phenomenon, so the DiamondView™ seeks answers in the other main distinguishing feature of synthetic diamonds, the internal growth structure of synthetic diamonds. The use of this as the point of differentiation on which the DiamondView™ is based serves only to re-enforce the importance of this property (and related features) in the separation of synthetic and natural diamonds.

The use of the ultraviolet fluorescence of synthetic diamonds to assist in their identification has been covered in Chapter 11. As we have seen, the presence of geometric patterning conclusively determines the synthesis of the stone. Use of this technique does however require skill, patience and a level of prior knowledge to enable effective interpretation of the results achieved. To this end, the DiamondView™ is designed to make the whole process simpler, and accessible to those without the underlying knowledge, or the time to perform manual gemmological tests.

The DiamondView™ is a larger and more visually elaborate instrument than the DiamondSure™. The instrument itself weights 20 kg and fits on a desk top, along with its accompanying specially configured computer. Special jaw style holders are available for both loose and ring mounted diamonds.[41] The component of the DiamondView™ instrumental in the detection of the synthetic diamonds works by illuminating the surface of the stone with a specialized ultraviolet light. This light is not of the wavelengths used by standard gemmological instruments (365 nm or 253.7 nm) but rather a wavelength shorter than 230 nm. This wavelength is used because it has been found that all types of diamonds (synthetic or natural) react very strongly to these shorter wavelengths of ultraviolet radiation, regardless of whether they fluoresce to the normal long wave or short wave UV light.

Why do all diamonds fluoresce to the wavelengths shorter than 230 nm? It is related to the band-gap theory mentioned briefly in Chapter 7. Those wavelengths shorter than 230 nm are equal to, or greater than, the intrinsic energy band-gap.[41] The fluorescence is also confined to the surface of the stone, which ensures a well defined, clear and sharper two-dimensional image of any fluorescent patterns present in the stone. Once stimulated, the fluorescence is viewed by a CCD (charge-coupled device) video camera combined with a variable magnification lens, and a visible light source for focusing. The image once taken is stored in the camera, and with a test time varying between 40 milliseconds and 10 seconds (depending upon fluorescent intensity), it returns a quick response to the computer.[41]

The stone can be rotated to enable viewing of different angles and facets. The use of a gear mechanism on the stone holder makes it unnecessary to remove the stone from the DiamondView™.[41]

The minimum requirements of the IBM compatible computer, enabling it to accept the images and display them in a reasonable time and with good clarity, are a 120 MHz Pentium processor (relatively small by today's standards), 32 Mb of RAM, PCI video input and graphics display cards. The software used for the DiamondView™ application is compatible with Microsoft® Windows™ version 3.1.

Hardware requirements include a high-resolution monitor in the order of 768–1024 dpi. The layout of the software is user friendly and the images are clear. These images of the internal growth structure of the diamond can be saved on the computer's hard drive to enable future viewing with ease. The number of images that can be stored is determined by the capacity of the computer's hard drive.[41]

Once the image is taken, and imported automatically to the computer, the image is displayed on the screen, enabling viewing of the patterns and colours of fluorescence followed by an easy analysis on the part of the operator. However, as the operator is required to interpret the results, the number of stones that can be effectively tested in a given time is significantly less than with the DiamondSure™.[41]

During the testing of the DiamondView™, 150 randomly chosen natural diamonds were tested together with the same 98 synthetic diamonds previously used to test the DiamondSure™. Concentric octahedral growth was evident (with blue fluorescence) in the natural diamonds tested,[41] with one particular natural yellow-brown diamond showing parallel growth bands formed by plastic deformation decorated by H3 defects. These are slip bands along which the diamond has undergone a shearing displacement. These defects are responsible for the 503 nm line in the spectra of Brown series natural diamonds.[41]

The synthetic diamonds exhibited a yellow to green fluorescence with clear geometric patterning resulting from the cubo-octahedral growth. The instruments also registered phosphorescence, which proved to be strongest in the octahedral zones.[41] Armed with such information, the professional gemmologist should have no trouble in conclusively separating natural and synthetic diamonds.

The DiamondSure™ is a relatively inexpensive device at a potential cost of a few thousand dollars. The DiamondView™ is more complex, significantly more expensive and is expected to be in the vicinity of tens of thousands of dollars.[41]

SAS2000 Spectrophotometer Analysis System

The SAS2000 Spectrophotometer Analysis System is based on a fibre-optic spectrophotometer with an optical resolution (half height/full width) of approximately 1.5 nanometres. It is integrated with users' existing computer equipment. Wavelength quantization is approximately 0.33 nm with an operating range of approximately 380–850 nm (other operating ranges available on special order). This is approximately eight times finer than the next closest single purpose colorimeter. The SAS2000 is a *dual* channel system which provides for more accurate tracking of any variations in the illumination source spectra. This is in contrast to the single channel colorimeters available today.

The SAS2000 is helpfully user friendly, and has been adopted by many professionals around the world. Distributors Adamas Gemological Laboratory recommend an initial warm-up for stabilization of optical benches of approximately 15 minutes. The sample diamond is then placed inside a three-inch integrating sphere on the sample pedestal. To prepare the instrument for the all-important sample scans it is essential first to take both a dark count and a white reference set of scans (31 scans are usually taken to minimize noise effects). These are performed with no illumination in the sphere, and again with the sphere illuminated but without the sample positioned. However, the white reference is taken with the gemstone in the sphere to account for 'substitution' error. Next the diamond to be tested is positioned over the sample probe to take a sample set of scans performed. Each set of scans takes approximately 30 seconds. Assuming that the scan integration time is not changed, only the white reference and sample sets of scans needed to be repeated for subsequent samples.

Unlike the DiamondSure^{TM} and the DiamondView^{TM} the system SAS2000 is more flexible and can be utilized for more than just identification of synthetic diamonds. As the system is calibrated with GIA GTL and HRD colour master (natural) diamonds (see Figure 11.2) it can be used for colour grading of diamonds (potentially synthetic or natural). Adamas Gemological Laboratory claim that the SAS2000 provides the most accurate colorimetry available for diamond colour grading in the market today. It produces a consumer oriented, copyrighted Diamond Quality Analysis Report offering conclusive evidence of assigned colour grades.

Confronted with a fancy coloured stone, the system can also provide origin of colour reports. The SAS2000 is programmed to automatically test the input data set for evidence of radiation damage at *a priori* wavelengths. Attention of the operator is then drawn to these potential damage sites. According to Martin Haske of the AGL, to date, the SAS2000 has not failed to report appropriate radiation damage in diamonds. The results for these particular irradiated diamonds have been verified by independent sources. These results were achieved at room temperature, although the diamond may be pre-chilled with liquid nitrogen to improve the resolution of the absorption lines.

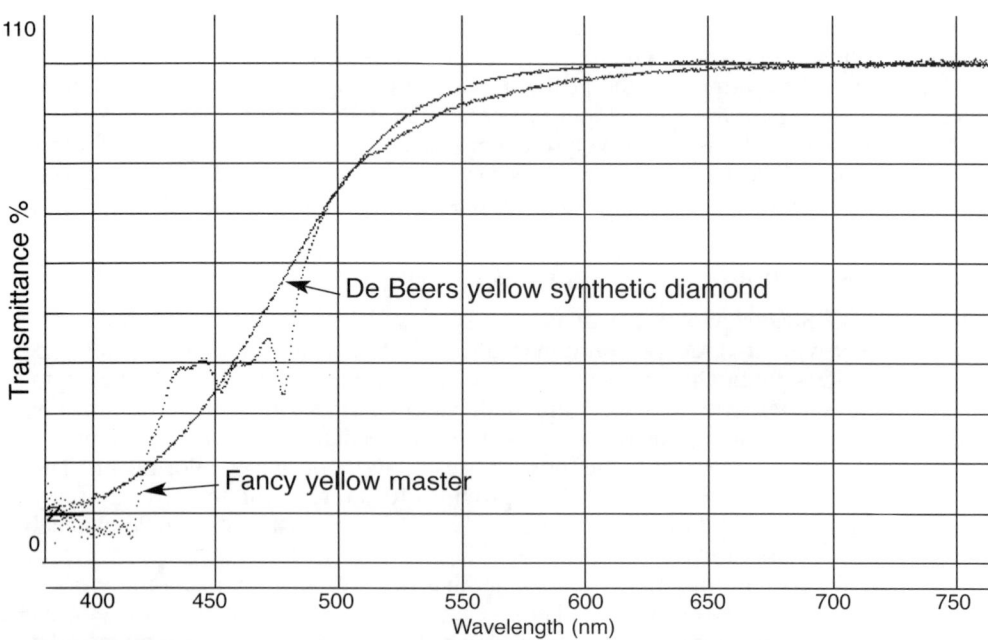

Figure 12.2 SAS2000 transmittance spectra of De Beers yellow synthetic diamond compared with the transmittance spectra of a fancy yellow natural colour master stone. (Courtesy Adamas Gemological Laboratory © 1998.)

The importance of the SAS2000 in this context, is its use in the identification of possible synthetic diamonds. The best approach in explaining the use of the system to this end, is to incorporate actual data. Figures 12.2 to 12.6 are examples of the SAS2000 spectrophotometer transmittance spectra for De Beers yellow, near-colourless and experimental blue synthetic diamonds. The results for the yellow and near-colourless synthetic diamonds are compared directly with the results for natural

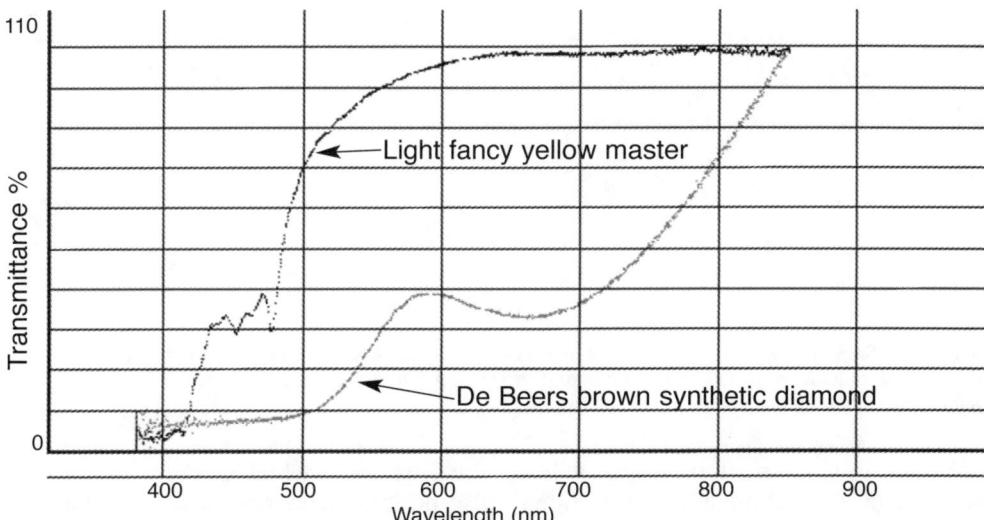

Figure 12.3 SAS2000 transmittance spectra of De Beers brown synthetic diamond compared with the transmittance spectra of a light fancy yellow natural colour master stone. (Courtesy Adamas Gemological Laboratory © 1998.)

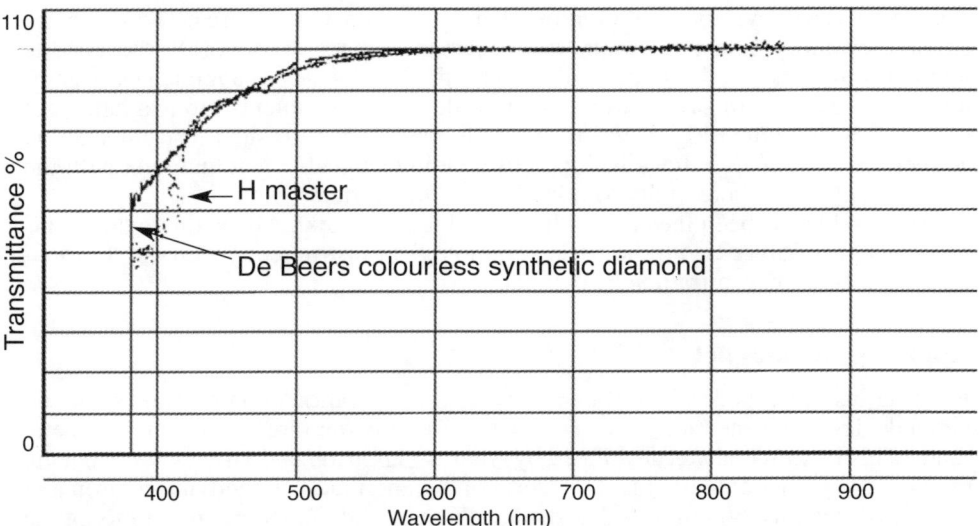

Figure 12.4 SAS2000 transmittance spectra of De Beers near-colourless synthetic diamond compared with the transmittance spectra of natural colour master stone graded as H. (Courtesy Adamas Gemological Laboratory © 1998.)

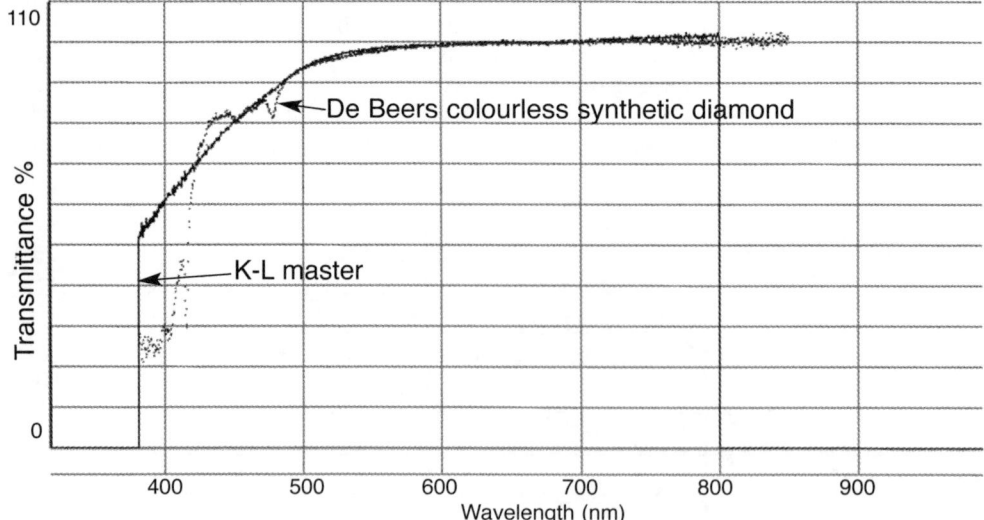

Figure 12.5 SAS2000 transmittance spectra of De Beers near-colourless synthetic diamond compared with the transmittance spectra of a natural colour master stone graded as K–L. (Courtesy Adamas Gemological Laboratory © 1998.)

diamonds of similar colour. It is easy to see the differences which can offer an excellent means of identification. Traditional absorption spectra can also be obtained through another mode of the same system. However, we have used the transmittance model here as the resulting data is clear, well spaced and the comparison more effective.

As suggested by Richard T. Liddicoat (editor-in-chief of the American journal *Gems and Gemology*), the invention of such devices provides a great service to the gem and jewellery industry. In our 'fast paced and competitive world', the trend is moving towards the need for such quick and reliable 'black-box' approaches to gemmology. He goes on to say, however, that while such instruments are available, not every practising gemmologist will realistically have access to such instrumentation, and there is no substitute for a well practised, knowledgeable gemmologist with the basic instruments and creativity to think beyond the norm.[78]

Certainly this has been the case with the author of this text. All synthetic diamonds procured were analysed with the basic gemmological instruments, with the information contained here as a frame of reference.

Exploiting the magnet

Two other instruments designed specifically for the separation of natural and synthetic diamonds deserve a mention here. The first of these is the rare-earth iron 'Magnetic Wand'. This instrument was developed by Alan Hodgkinson and utilizes the magnetic attraction of synthetic diamonds. It consists of a wand made of neodymium, iron and boron, mounted on a 60 mm wooden shank. When brought into close proximity of the stone under test, it induces rotation or movement in a suspended synthetic diamond.[79]

It is 5 mm in diameter and has been found to attract Sumitomo, De Beers, and

Specialized separation instruments 113

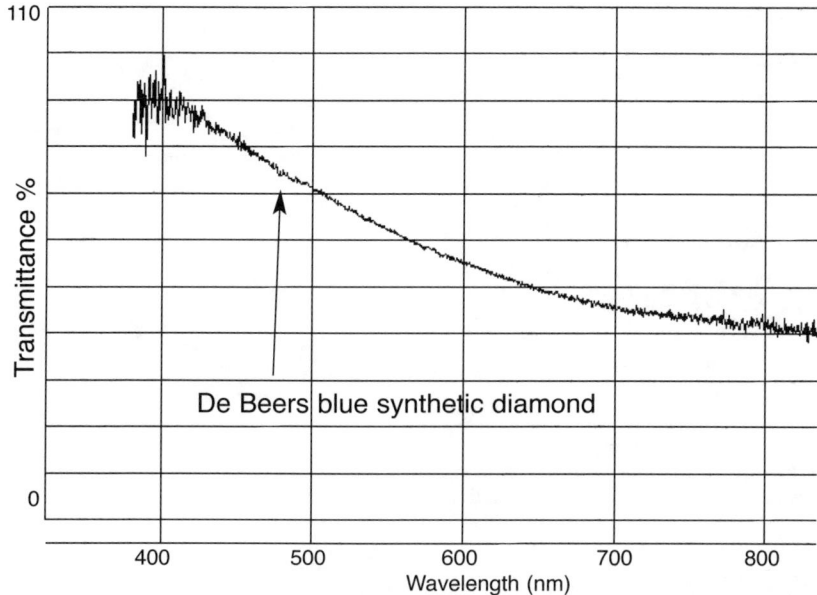

Figure 12.6 SAS2000 transmittance spectra of De Beers blue synthetic diamond. (Courtesy Adamas Gemological Laboratory © 1998.)

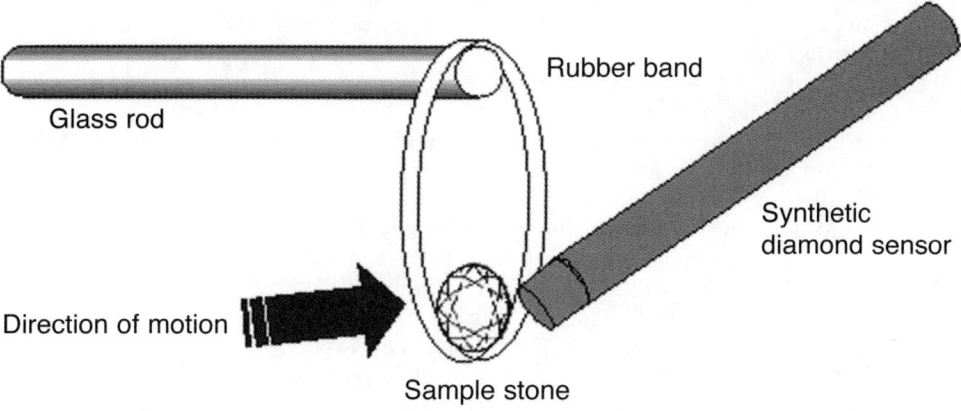

Figure 12.7 Schematic diagram of synthetic diamond suspended in a rubber-band 'cradle'. The arrow indicates the expected direction of motion.

Russian synthetic diamonds. The Magnetic Wand is very powerful, said to be the strongest compact magnet currently available, and should be kept away from sensitive electrical equipment or magnet information strips and film.[37]

Along the same lines is the Linton Synthetic Diamond Sensor. This affordable instrument combines the strong neodymium-boron-iron magnet in a light-weight and easy-to-use plastic case. Tests conducted by the author found this sensor to be very effective. There are a number of methods for conducting experiments of this type, depending upon the size and shape of the diamond in question.

Large diamonds may be suspended in a nylon or cotton thread arrangement;

114   The diamond formula

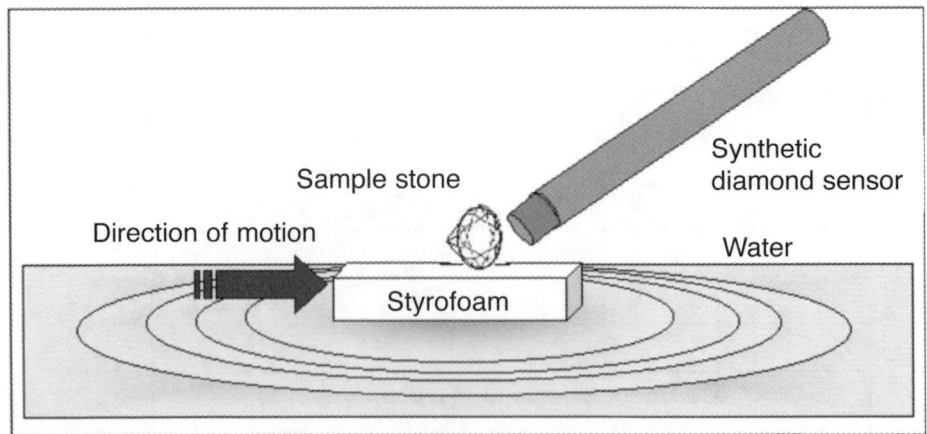

Figure 12.8 Schematic diagram of synthetic diamond mounted on a styrofoam raft in a water bath. Arrow indicates the expected direction of raft motion.

however, it is essential not to use Blu-tack® as this substance is also magnetic! Keep the suspended stone out of any draft and ensure that your own breath does not disturb the stone. Bring the sensor, or wand, in close proximity to the diamond and watch for movement toward the magnet. A thick rubber band may also be useful as a basket in this instance.[72] Consider the configuration depicted in Figure 12.7. Suspend the rubber band by a glass or plastic rod, as these are non-magnetic. By viewing this arrangement laterally and aligning the rubber band with a vertical line of reference, one can easily observe motion in the horizontal direction resulting from the attraction of the diamond to the magnet.

Large stones may also be tested by placing them on angled glossy paper, such that the stone almost slips down the paper, but not quite. Do not place the stone table down. Bring the sensor/wand close to the lower side of the stone, and watch for movement.[79] Small diamonds or stones already set in jewellery may, however, be difficult to suspend in such an arrangement, or they may be too light as to induce movement.

To test these stones other methods must be employed.

The first and easiest is to place the sensor/wand in contact with the stone to see if the diamond sticks to the magnet when the wand is elevated. Ensure that the magnetic tip is clean and any attraction is not due to contamination of sticky substances on the stone or magnet. Second, small diamonds can be placed on a raft composed of styrofoam or polystyrene, floated in a dish or water. Use the minimum amount of styrene possible in constructing the smallest raft that will still float the diamond under test.

Place the stone pavilion down on the raft, and orient the raft carefully in the centre of the dish. Wait until the water is still. Bring the sensor/wand close to the table of the stone and watch carefully for movement. Surface tension can also have the effect of moving the raft to the sides of the dish, so repeat this experiment to verify any results. Additionally, to reduce the effects of surface tension, it is a good idea to add a few drops of liquid detergent to the water.

This method can be modified to be suitable for set stones, by using a larger raft and submerging the setting through the raft and into the dish of water below. With just the stone protruding over the surface of the raft, the method is the same as above. Precious metals are not magnetic, however they are heavy and for this reason the suspension method is not suitable.

All of these methods were tested by the author and found to be effective.

Chapter 13

Treated synthetic diamonds

Treatment of gemstones is widespread in the gem and jewellery industry. Some treatments are accepted, and others rejected. The diamond industry is no different. Laser drilling and fracture filling of diamonds are better known than other treatments that are possibly less obvious, but just as effective.

Irradiation of diamonds has been practised for decades. The methods and techniques have been refined to a high degree with many texts offering detailed synopses on the subject. Available texts on the treatment of gemstones include *Synthetic, Imitation and Treated Gemstones*, O'Donoghue, 1998. We will bypass a full description in favour of brief explanations during discussion on the treatments identified with relation to synthetic diamonds.

Detailed reports exist of two such treatments that have potentially important repercussions for the industry. These are the colour-treatment of two synthetic diamonds submitted to the GIA Gem Trade Laboratory, and treated Russian synthetic diamonds, again recorded by the GIA.

Seeing red

In July 1993 the GIA Gem Trade Laboratory received a red diamond from a Bombay diamond dealer for routine origin of colour report. In September of the same year another red diamond was submitted to the laboratory in New York, again for an origin of colour analysis.

Red diamonds are beautiful, rare and extremely expensive. Few diamonds have been officially assigned the colour red over the years, for good reason. There are vast amounts of money at stake. Descriptions of colour in diamonds have varied names, some fanciful, others straightforward. This is apparent in discussion of the rich and vibrant colours of the top fancy coloured diamonds. Red is among the most prized of them all.[80]

The owners of these red coloured diamonds had much depending on the results of the GIA reports. The official stamp of approval from the laboratory, and the assigning of the red colour description, could mean huge rewards for the owner as natural red diamonds can attract record market prices. Unfortunately for the owners of these diamonds, this was not to be. Instead their valuable gems offered up new and as yet undiscovered secrets.

In short, these diamonds were not naturally red. In fact, they were not natural at all. The GIA found that the diamonds were in fact synthetic and the red coloration was induced artificially. This was the first such case submitted by a member of the public. By this stage other synthetic diamonds had been submitted to the laboratory. Staff were not unfamiliar with the situation, and consequently were not unprepared. The identification of the synthetic origins was readily made using standard gemmological equipment. More advanced analytical tools were also used, no doubt due to the potential value of the 0.55 ct round brilliant cut and the 0.43 ct radiant cut diamonds.[61]

The concept of treating synthetic diamonds is not new. Literature had been published with regard to pink and red synthetic diamonds in 1978[81] and 1991.[82] These references also stated that the colours of the stones referred to in the text were attributed to treatment by irradiation followed by annealing.

The GIA described the colours for the two diamonds submitted to them in 1993 as dark brownish orangy red, and dark brownish red respectively.[61] The instruments employed in the identification and verification of these synthetic diamonds included a standard gemmological microscope, both long and short wave ultraviolet light sources, prism and diffraction grating gemmological spectroscopes, spectrophotometer, infrared spectrophotometer and an EDXRF system.[61] The later technologies are discussed in more detail in Chapter 9.

Many of the results of these tests mirrored those expected for synthetic diamonds. Characteristic results were observed in the areas of microscopy, and patterning in ultraviolet fluorescence. However, in other areas these synthetic diamonds differed greatly from other synthetic diamonds, and from natural diamonds of similar colour. Ultraviolet fluorescence in these artificially red stones varied in a number of ways from the typical synthetic diamond results. First, for the 0.55 ct round brilliant cut synthetic diamond, two colours were seen under both long wave ultraviolet and short wave ultraviolet illumination, as an uneven distribution of a moderately intense reddish orange and a very intense green. This is significant as these colours are not usually seen in the fluorescence of synthetic diamonds, nor are multiple colours usually visible.[61]

The green colour appeared as a combination of distinctive cross and square shaped patterns, and continued to phosphoresce briefly after the short wave ultraviolet light was shut off. Scientists observed the reddish orange areas in different parts of the stone, but confined to a small area near the girdle of the stone under long wave ultraviolet, and throughout the stone under short wave ultraviolet light.[61] As Chapter 12 stated, synthetic diamonds rarely exhibit fluorescence under long wave ultraviolet light. This makes the case of this red synthetic diamond all the more unusual. But it was not alone in the manifestation of extraordinary ultraviolet fluorescence.

The other smaller radiant cut red synthetic diamond also revealed uneven and multi-coloured results. In the case of this 0.43 ct synthetic diamond, the fluorescence proved more intense under short wave ultraviolet, and moderately intense under long wave ultraviolet. The patterning expected of synthetic diamonds, on the vein of the geometric zoning mentioned previously, was present; however, the colour of the luminescence exhibited by this gem was even more unusual than the round brilliant cut synthetic diamond described above. Under both long and short wave ultraviolet, when viewed through the table of the stone, a small red area was visible, surrounded by a fine green fluorescent band. From the corners of the green fluorescent rim, orange luminescent bands extended toward the corners of the table of the diamond, meeting with stronger orange fluorescent square shaped zones. The remainder of the stone exhibited a less intense, orangy red fluorescence, apparently of even distribution.[61]

This becomes more credible when the distinct colour zoning seen within each stone, viewed under magnification, is taken into account. Both red synthetic diamonds exhibited light yellow square-shaped, and cross-shaped areas surrounded by the deep red areas. These yellow sections were perceived to be tabular and narrowed, in relation to the red areas.[61]

Spectroscopic analysis of the 0.55 ct red synthetic diamond revealed a number of sharp bands in the visible realm. Absorption bands around 500 to 660 nm were sufficiently intense to be visible through the hand-held gemmological spectroscope when the stones were cooled with liquid nitrogen. The 0.43 ct radiant cut synthetic diamond revealed a less outstanding spectrum. However, the lines that were present corresponded to similar positions as seen in the round brilliant cut stone.[61]

In addition to these absorption bands both synthetic diamonds exhibited increasing absorption toward the violet end of the spectrum and a broad region of absorption between around 500 nm to approximately 640 nm.[61] As we have seen in Chapter 12, synthetic diamonds rarely exhibit a spectrum. Many of these bands were attributed to nickel and nitrogen in the lattice (refer to Table 8.1), which were no doubt included through the catalyst. Advanced analysis (EDXRF) of the stones' chemical composition found the presence of nickel and iron.[61] Figure 13.1 shows the result of EDXRF spectra of these red specimens, with the peaks due to nickel and iron marked accordingly.

Metallic inclusions were visible as rounded opaque inclusions with a reflective metallic lustre in the 0.55 ct stone, and as unusual square shaped inclusions on the 0.43 ct stone along with a large open cavity. Minimal graining was present in both stones, marking the intersection of the coloured zoning. If these stones' clarity *were* to be graded, they would receive an 'Imperfect' grade.[61] Using infrared spectroscopy, scientists found that the diamonds were of mixed type.[61]

The characteristics described are significantly similar in temper to known charac-

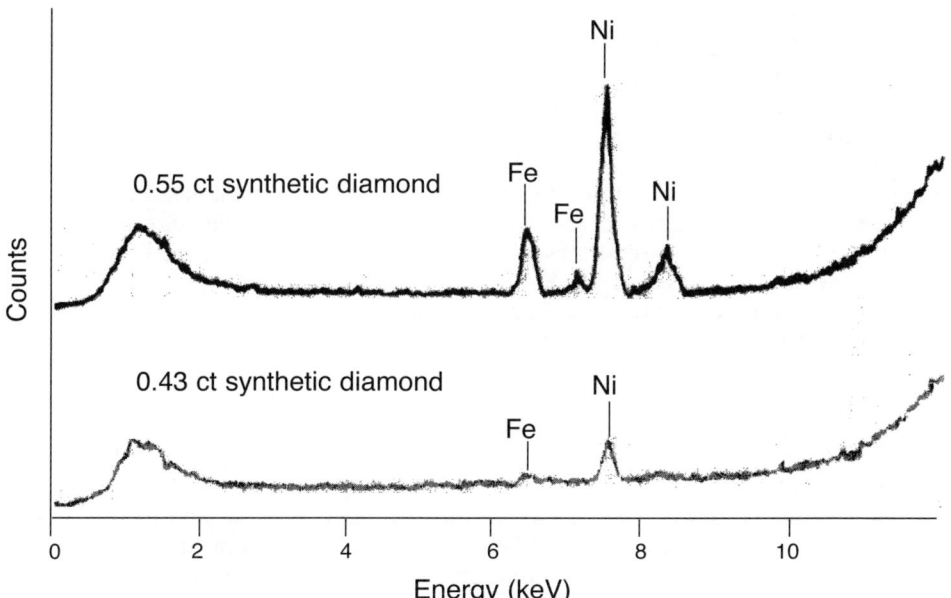

Figure 13.1 EDXRF spectra of two red synthetic diamonds indicating the presence of nickel (Ni) and iron (Fe) in the metallic inclusions. (Courtesy *Gems and Gemology* **29**, 3 © GIA 1998.)

teristics for synthetic diamonds for it to be determined accurately that these two red diamonds are indeed synthetic. The colour zoning, metallic inclusions and patterned fluorescence to ultraviolet light are all observed in synthetic diamonds, and parallels can be drawn from the results described in Chapter 12 for other synthetic diamonds. However, as stated by the GIA, there are also significant differences from the results obtained from previously studied synthetic diamonds.[61] The most obvious of this is the red colour, which is achieved by treatment.

Previously recorded synthetic diamonds have been inert to long wave ultraviolet light. This was not the case with these red synthetic diamonds. Similarly, synthetic diamonds usually lack a spectrum, while these specimens displayed sharp absorption bands and increasing absorption toward the violet. The most alarming results were the multi-coloured fluorescent patterns observed. By this time the GIA had some experience in treated synthetic diamonds, having seen orangy red Sumitomo synthetic diamonds, and these stones also displayed orangish fluorescence to both long wave and short wave ultraviolet light following treatment.[61] The green fluorescence seen in these red synthetic diamonds, however, is thought to be due to the formation of a H3 centre[83] in the diamond lattice. The orangish coloured fluorescence is believed to be almost certainly due to the treatment. This is in part because similar coloured fluorescence has been seen in treated pink natural diamonds.[61]

Treatment, in this case, can be identified by the spectral features observed by infrared and low temperature spectroscopy.[61] It is known that type Ib diamonds will change from a yellow colour to a pink or red colour following treatment by irradiation and subsequent annealing. This will occur in both natural and synthetic diamonds[84] and is believed to be the case with both of these synthetic diamonds. The 550 nm band found in the visible spectra for these diamonds is believed to be due to a defect produced by plastic deformation along a glide plane.[61] Essentially these diamonds are treated yellow type Ib synthetic diamonds. This serves to separate them from natural pink or red diamonds.

In fact although many of the results were more akin to those observed in natural pink or red diamonds than in other synthetic diamonds, all known pink to red natural diamonds can easily be distinguished by their type. Treated natural pink to red diamonds can also be distinguished by the lack of the zoning patterns. These synthetic stones had fluorescence attributed to the synthetic formation, and the nickel related spectral bands.[61] This is good news for the gem industry. As synthetics and treatments become more widespread it is of great advantage to know that detection methods are available, especially as many of these features are visible using standard gemmological equipment.

A more detailed analysis of these findings, especially the spectral features and their significance, can be found in the paper published by the GIA gem trading laboratory team, listed in the notes.

The Russian treatment

Much has been made in the press and journals of the intention of Russian synthetic diamond producers to sell their product to the jewellery industry, and therefore indirectly to the public. Chatham Created Gems has openly commenced the sale of synthetic diamonds, apparently derived from an undisclosed Russian source.

While the problem associated with the detection of synthetic diamonds in general has been discussed in Chapters 10–12, here we will investigate another aspect of the Russian synthetic diamonds and the additional problems it may pose. Not only are the synthetic diamonds produced by a different method (as outlined in Chapter 4), but following the synthetic procedure, some of these diamonds are treated.

The GIA in 1993 had the opportunity to study some samples of treated Russian synthetic diamonds, along with some as-grown samples, in order to enable a comparative framework. Little other information is available on the subject at this time, therefore most of the information here is related to this study and the observations made by the GIA scientists. The initial report to the GIA was that the synthetic diamonds had been 'annealed', or heat-treated in a high-pressure environment after synthesis (HPHT treatment).[33] The data compiled with regard to the treated samples is all taken post-procedure, and, as no data were taken before the treatment, the only comparison that could be made was with other samples. This means that to a certain extent one cannot be sure of the changes, if any, in properties. It is however safe to say that the treated diamonds varied in a number of ways when compared with the untreated samples supplied at the same time and believed to be the same type of synthetic diamond, made in the same BARS apparatus.[33]

In all, the GIA were supplied with 10 yellowish Russian synthetic diamonds, seven of which represented the as-grown state, and three of which had undergone the annealing treatment. All samples were produced in Novosibirsk, a major city in southern Siberia approximately 3000 km east of the capital Moscow, famed for its technologies. It is interesting to note that these samples were consistent with other Russian synthetic diamonds in many respects, indicating that previously observed specimens may have been produced in the same way.[33]

The as-grown synthetic diamonds consisted of two crystals and five faceted gems, ranging in colour from yellow or orangy to brownish-yellow. The treated synthetic diamonds were all faceted and were of a greenish-yellow colour.[33] The sizes of the samples also varied from 0.88 ct (for an as-grown sample) to 0.11 ct (also an as-grown sample), with all other samples falling within these weight parameters. During the study, standard gemmological instruments were used, with the addition of an infrared spectrophotometer, EDXRF system and Luminoscope.[33]

On average, it was found that these synthetic yellow Russian diamonds 'exhibited greater variation in their gemmological properties' when compared to other previously tested Russian synthetic yellow diamonds.[33] As the majority of the samples were faceted we will omit information relating to the comparison of crystal morphology in this case. Our attention will be focused on the characteristics effected by the annealing treatment, and on the information known about the treatment itself.

Research by Russian laboratories on the subject of treatment was originally undertaken in order to investigate the effects of the annealing, and were employed using the BARS high-pressure, high-temperature device used in the initial synthesis. The reported conditions of the annealing treatment involved exposure of the sample to temperatures from 2000°C to 2200°C, at pressures of 70 000 to 80 000 atmospheres, for around four to five hours. The main effect of this treatment seems to be on the colour of the synthetic diamond and the spectra, both in the visible and infrared range.[33]

This treatment manifested in more than just the observed greenish hue attributed to the treated diamonds. The hue was absent from the as-grown specimens. All faceted samples exhibited colour zoning related to the internal growth sectors of the synthetic diamond. These zones did however appear more conspicuous in the treated specimens than in the as-grown synthetic diamonds. Naturally the zoning varied in relation to the cut of the stone, and the viewing conditions. However, the zoning within the treated synthetic diamonds proved more consistently noticeable when viewed in different conditions and from different directions. As expected, the pattern and orientation of the zoning corresponded to the ultraviolet luminescent pattern and graining.[33]

Graining was also observed in the Russian synthetic diamonds, including surface and internal varieties. One of the treated synthetic diamonds displayed rather prominent graining in the form of a bevelled square shape (octagon) with radiating arms on the surface of the table facet.[33] This same specimen also displayed internal graining of a similar nature. This appeared as a rectangular shape with graining lines emanating from each corner.[33]

No chemical differences were detected between the as-grown and treated samples following analysis with EDXRF. Surprisingly little stain was detected in any of the faceted synthetic diamonds (both treated and untreated) under polarized light, with no differences reported there either.[33] The other differences were detected in the ultraviolet luminescence and the absorption spectra.

One anomaly observed with these Russian synthetic diamonds was the luminescence to long wave ultraviolet light. Only one of the as-grown samples was inert to long wave ultraviolet, a property usually attributed to *all* synthetic diamonds.[33] As expected, all of the as-grown samples fluoresced to short wave ultraviolet radiation, either equal to or exceeding that of the long wave ultraviolet reaction.[33]

The treated synthetic diamonds exhibited slightly different fluorescent phenomena. All three specimens fluoresced more intensely to long wave ultraviolet than to short wave ultraviolet. There was the unusual addition of moderate to strong phosphorescence observed following exposure to both long and short wave ultraviolet. It is of interest that it is only the treated synthetic diamonds that phosphoresced. This lasted for approximately 30–45 seconds for long wave ultraviolet and for a period of 30–60 seconds for short wave ultraviolet. In all cases, long wave ultraviolet and short wave ultraviolet, for both the as-grown and treated samples, zoned patterning was reported. This was consistent with the patterns manifest in the colour zoning and graining. The luminescence was, in all cases, variations of a greenish-yellow or yellow colour. This greenish-yellow fluorescence was also indicated by the luminescent spectra of the treated samples.[33]

In addition to this it was found that the luminescent colour remained consistent even when the wavelength of the stimulating radiation was modified, but the intensity for long wave increased around 365 nm and for short wave decreased around 255 nm.[33]

The visible range absorption spectrum was observed by using a hand-held spectroscope, and the samples were cooled, using a spray refrigerant. As we have seen in Chapter 12, type Ib synthetic diamonds (accounting for the yellow synthetic diamonds) rarely have a spectrum. This is not unexpected because a spectrum is not exhibited by natural type Ib diamonds either! All but two of these Russian synthetic yellow diamonds displayed absorption bands in their respective spectra. The exact positions of these bands were then determined using a spectrophotometer; this subsequently revealed even more bands.[33]

The as-grown samples each exhibited increased absorption toward the violet end of the visible range, beginning at approximately 450–500 nm. The bands distributed throughout the spectrum of the as-grown synthetic diamonds were found to differ from those observed in the case of the treated synthetic diamonds. The treated stones exhibited much less absorption below 450 nm, and a number of bands not present in the spectrum of the as-grown specimens. The positions of these bands are outlined in Table 13.1. The majority of these bands were attributed to the presence of nickel and

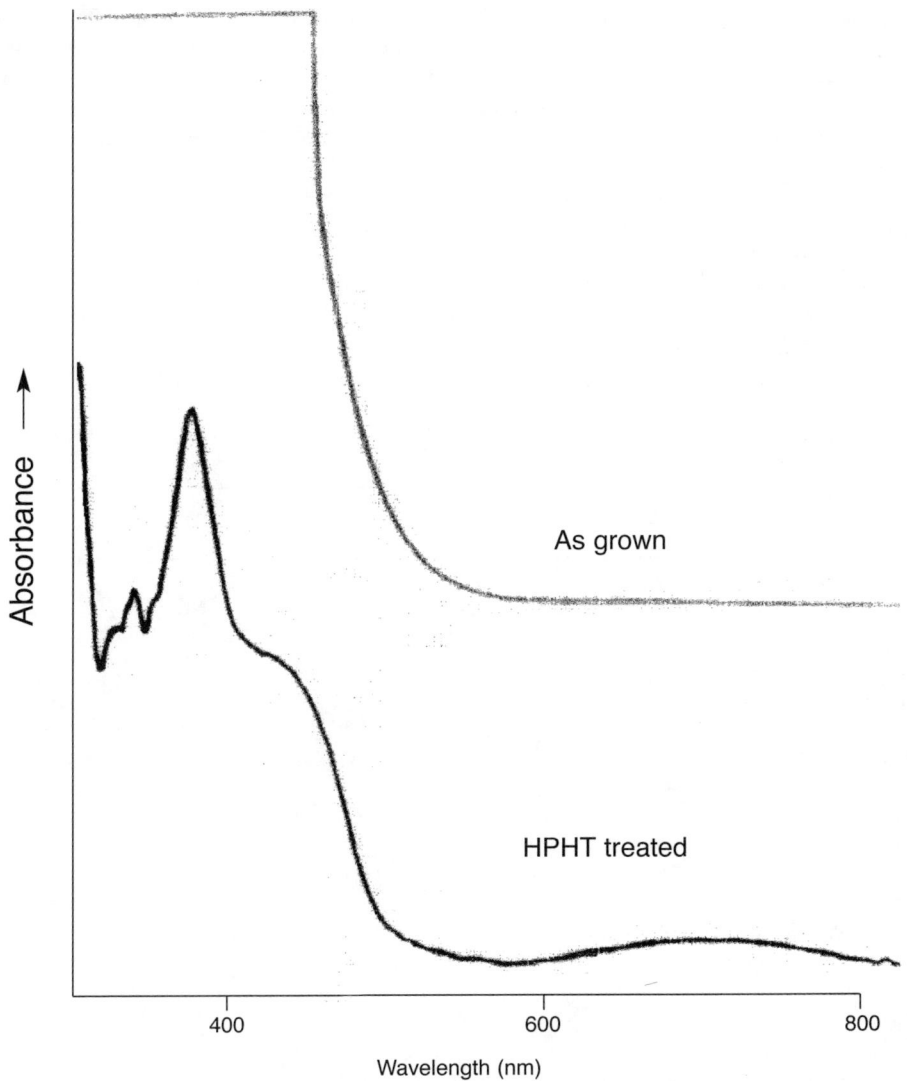

Figure 13.2 Room-temperature absorption spectra of an as-grown Russian synthetic diamond (top) with increasing absorption toward the ultraviolet, and a HPHT treated Russian synthetic diamond (bottom) with bands indicated by peaks in the violet region of the spectrum. (Courtesy *Gems and Gemology* **29**, 4 © GIA 1998.)

nitrogen.[33] Figure 13.2 shows the visible range spectrophotometer absorption data from a yellow treated and an orangy-yellow as-grown sample at room temperatures.

The spectra has been offset vertically to enable easy comparison. The upper result is that of the as-grown synthetic diamond and the lower corresponds to the treated synthetic stone. Note the increasing absorption toward the violet end of the spectrum (left), but the decrease in absorption below 500 nm in the treated stone. The broad absorption region between 600 nm and 800 nm accounts for the greenish component in the colour observed in some of these Russian synthetic diamonds.

Overall the spectral bands of the three treated synthetic diamonds were described by the GIA team as 'quite distinctive'. There were more bands in total and yet less extinction toward the violet. It was deduced that the process of treating the diamonds also had the effect of increasing the transparency of the stone in the 300 nm to 500 nm range. The bands centred mainly around the shorter wavelengths, and extinction, in the case of the as-grown synthetic diamonds, also explained the yellow body coloration. The greenish yellow colour in some of the samples was found to be due to a broad and rather dominant band in the shorter wavelengths of around 700 nm visible in the spectra of the stones.[33]

Table 13.1 Variations in the spectra lines observed in as-grown Russian synthetic diamonds and HPHT treated Russian synthetic diamonds.

Wavelength (rounded to nearest nm)	As-grown synthetic diamonds	HPHT treated synthetic diamonds	Visible with hand-held spectroscope
792 nm	✓	✓	No
767 nm		✓	No
732 nm	✓		No
711 nm	✓		No
691 nm	✓		No
671 nm	✓		No
660 nm	✓	✓	No
658 nm	✓		Yes
649 nm	✓		No
647 nm	✓		No
639 nm	✓	✓	No
637 nm	✓		Yes
627 nm	✓		No
623 nm		✓	No
617 nm	✓		No
607 nm		✓	No
592 nm		✓	No
582 nm		✓	No
563 nm		✓	No
553 nm	✓	✓	Yes
547 nm	✓	✓	Yes
540 nm		✓	No
527 nm	✓	✓	Yes
520 nm		✓	No
518 nm		✓	Yes
516 nm	✓	✓	No
511 nm	✓	✓	Yes
506 nm		✓	No
503 nm		✓	Yes
500 nm	✓	✓	No
494 nm	✓	✓	No
491 nm		✓	No
481 nm		✓	Yes
478 nm		✓	Yes
477 nm		✓	Yes
473 nm	✓	✓	Yes
468 nm	✓	✓	No

The main spectral features were categorized by the GIA scientists and the results of the as-grown and treated synthetic diamonds compared. The reference contained within the notes is recommended to those interested in a more detailed analysis of these spectral features and their categorization.

Infrared spectroscopy found that these Russian synthetic diamonds were in fact formed of mixed type. The addition of the type Ia to the usual type Ib was very unusual at the time. It was determined that the treated stones contained type IaA aggregates that came about during a conversion process during treatment.[33] The manifestation of mixed type synthetic diamonds is discussed in more detail in Chapter 14.

The GIA scientists summarized the effects of the Russian treatment on the synthetic diamonds and the way it was manifested in the results observed. These Russian stones were the first synthetic diamonds treated in this way to be studied by the GIA Gem Trade Laboratory. While the application of this treatment can increase the challenge in identifying synthetic diamonds, it identifies additional gemmological properties, including the low temperature spectroscopy (visible with the hand-held gemmological spectroscope).[33]

Results of studies on treated diamonds had been reported prior to this exercise,[85] as had the existence of this type of nickel related spectral phenomenon.[86] This topic has become more important due to the sale of Russian synthetic diamonds (in particular) to the jewellery industry. In general the Russian synthetic diamonds produced using the BARS device were found to have unique properties that effectively added to the repertoire of synthetic diamond identification.[33]

Reference has been made during this chapter to some more advanced scientific observations, and the instruments used. We will now take a more detailed look at these techniques and their relationship to synthetic diamonds.

Part 3
The New Diamond Makers

Chapter 14

CVD and the new breed of diamond makers

We have discussed the fact that diamond is the hardest naturally occurring substance, and why this is the case (see Chapter 3), and it is this very property on which many industrial applications are based. Imagine the future of the industry if other objects could be as hard and long wearing as diamonds. Imagine eyeglasses that never scratch, crash helmets that resist impact cracking or the hardness of diamond on flexible surfaces. This may be the future of diamond coating and diamond thin films.

In Parts 1 and 2 of this book, we have addressed the invention, development, process and industry of high-pressure diamond synthesis. The trouble associated with the application of high pressures, and the expenditure in both technology and time, have been vast. It may come as a surprise to some to find now that diamond can also be formed in an environment with a pressure close to zero. Amazingly these are the conditions under which diamond films and coatings are produced. This goes against everything we have learned about the synthesis of diamonds; how can it be so? To solve this paradox we must first delve into the history of the creation of diamond thin films, and then unlock the secrets of the new breed of diamond makers.

Low-pressure diamond synthesis

Low-pressure diamond synthesis was the dream of scientists of the 1800s. To make the most precious of gems at ambient pressures, under the same conditions as we live and breathe, was a dream that for many years seemed practically impossible. It was known that diamonds formed at great depths, where the pressure was intense. Early diamond making efforts focused very heavily on the application of very high pressure in an attempt to replicate this. Was this all in vain? Could it be possible to make diamonds at room pressure? Or lower pressures, such as in a vacuum? It was not until the early 1900s that one scientist dared to seek answers to these questions.

Scientists have now learned that the history of low-pressure diamond synthesis potentially pre-dates mankind. This kind of diamond formation has been occurring for billions of years and the resulting diamond could be infinitely common, though not on our planet. Deep in space in the veritable voids between the stars and clouds of interstellar dust, a near vacuum exists. It is here that hot carbon atoms find themselves attracted to other carbon atoms. This results in the atom bonding to four carbon neighbours. The result is tiny molecules of interstellar diamond dust.[2] The interesting point that became the stumbling block in the low-pressure synthesis of diamonds was that, for reasons unknown at the time, carbon atoms seem to prefer to bond with four more carbon atoms rather than the three carbon atoms forming the thermodynamically stable state of graphite. This four-atom bond was that required for the tetrahedral diamond lattice.

It was long suggested by scientists that diamonds could be synthesized at pressures at which graphite was the thermodynamically stable form.[2] The first recorded incident of synthesis of a thin diamond coating produced in the laboratory was in 1911, when W. Von Bolton reported to have experimentally coated a natural diamond seed crystal with a man-made diamond layer. He achieved this by using acetylene gas in a mercury vapour at ambient pressure and temperature of 100°C![87] His experiment lasted around two weeks and was generally regarded as an academic experiment for the sake of curiosity and no further interest in the field was generated.

It was not until the 1950s that interest in the concept of low-pressure diamond synthesis was rekindled. Bridgman discussed this possibility in 1955.[6] Two scientists located on the other side of the globe, in the USA and the former USSR, simultaneously re-analysed low-pressure diamond film technology. The Russian scientist Derjaguin and his soviet team in Kiev began work in the early 1960s and eventually had success in 1968.[88] This team was able to grow faceted diamond crystals in a vacuum, a truly amazing result at the time.[2] He was not the first to achieve success (that honour went to Union Carbide), but it is Derjaguin's method that formed the basis for modern technology.

At the time, however, there was much speculation in the western scientific community as to whether the films Derjaguin had produced were in fact diamond.[35] To further fuel speculation, Derjaguin also claims to have deposited an author's certificate regarding his process as early as 10 July 1956, in the Soviet Union.[6]

American W.G. Eversole, working for the Union Carbide Linde Air Products Division, and Turnbull and Oriani at General Electric, were among the first to re-analyse the potential applications for diamond coatings and conduct research into the area within the United States. The teams worked in parallel, in an environment reminiscent of the ASEA and General Electric Project Superpressure, which was also being conducted in the 1950s. The General Electric low-pressure diamond synthesis team received strong corporate backing, but were surpassed (without any successful progress) in 1957 when the high-pressure Project Superpressure team made their revolutionary breakthrough.[2]

W.G. Eversole succeeded and in 1958 was awarded two US patents in 1962. His patents (numbers 3,030,187 and 3,030,188) describe the process of passing carbon-containing gas over a seed crystal under correct conditions to enable deposition of diamond carbon on the seed.[89] At best the material only increased in weight by around 2 per cent per hour, which, although heralding a new era in diamond synthesis, was not nearly as glamorous as the results achieved by high-pressure diamonds synthesis techniques.[2]

While these appear to be the first patents in the field of low-pressure diamond synthesis, this is not the technique reminiscent of today's methods. This method was however reconstructed six years later, following the granting of a patent to researchers at Case Western University, in Cleveland, Ohio. Here Angus (now a leader in the field),[90] Will and Stanko were able to deposit new diamond on meticulously cleaned natural diamond powder at a temperature of 1050°C, and less than a thousandth of an atmosphere. This discovery proved beyond a doubt that diamonds could be grown in sub-atmospheric conditions. Following this achievement it was left to another international community of scientists to make the next step in low-pressure diamond synthesis.[2]

Next it was Japan's turn to lead the way. Japanese researchers S. Matsumoto and S.

Sekata at NIRIM laboratories produced entire discs of synthetic diamond by low-pressure techniques in the late 1970s and the early 1980s.[2] Serious research commenced in 1974 at NIRIM in Tsukuba.[1] This was built on the previous work of Derjaguin and his Russian team, and proved that this method of deposition of diamond thin films at low-pressure was technically possible.[91]

However the Japanese went further, improving the technique by the addition of hydrogen into the chemical cocktail.[92] In 1981 the Japanese team published in English a document which served as a veritable catalyst to the worldwide research community. This was a document that saw diamond film technology become the focus of many institutes around the globe.[1] In 1988 around US$20 million was spent on CVD (chemical vapour deposition) research in Japan alone.[35]

The achievements of the Japanese rekindled a new interest in low-pressure diamond synthesis in the west. R. Roy from Pennsylvania State University visited Japan in 1985 and was instrumental in the formation of a research consortium (combining universities and commercial companies) devoted to progress in this field. They realized that the current technology used to manufacture silicon-based superconductors (of which computer chips are built) was also suited to diamond formation.[2]

Today joint ventures in a number of companies, universities and government departments involved in commercial production of diamond thin films,[35] born from this flurry in the 1980s, have further perfected methods. Research still continues into decreasing the temperatures required, and into making the process more cost effective.[2]

How then are diamonds synthesized in a low-pressure environment? Why does the carbon form diamond and not graphite? That is the subject of our next section.

The magic of low-pressure diamond synthesis

Diamond film studies are complex, extensive and continually evolving. We cannot hope to cover all aspects of this science, nor do justice to all facets of the area. Indeed, diamond films and their manufacturing processes are the subject of many books in their own right. Here we will confine this discussion to a gemmological perspective, to delve into the secrets of diamond films and coatings in order to explain the nature of the medium.

The process for creating synthetic diamond films or coatings is substantially different from that adopted in the synthesis of diamond grit or mono-crystals. In the study of high-pressure synthetic diamond grit and mono-crystals we analysed each of the necessary elements needed for success. Here we will briefly do the same to be consistent, and to maintain our continuity.

As some of you may have deduced from the history of low-pressure diamond synthesis, graphite is not the source material; neither are any of the other suggested solid-state carbonaceous materials. As a source, this process involves the deposition of carbon from the gas phase. The gases most commonly used are methane (CH_4) or propane (C_3H_3).[6] They are called hydrocarbons. Other gases with similar chemical composition are principally carbon and hydrogen. Hence from here on we will refer to hydrocarbons, rather than specifying methane or propane (or other hydrocarbons).

These gases are composed of molecules of the hydrocarbon interacting in a free and random way. This means that these gases are free to decompose or bond in a stable fashion. They may form graphite or diamond (or amorphous carbon), depending upon the conditions of exposure. These conditions, just as in the case of high-pressure diamond synthesis, are set by the pressure of the environment and the temperature, or

130 The diamond formula

energy level applied. Here more differences in the two types of diamond synthesis are evident.

The element of pressure represents another major contrast, a substantial step away from traditional diamond making. As previously mentioned, traditional diamond synthesis has meant high pressure, which constantly ruptured anvils and devices. This is not the case in this new and exciting methodology. Here the need for high pressure, and all associated problems, is effectively removed from the equation. Diamond thin films and coatings under discussion are produced in low-pressure conditions. The range is from atmospheric, near-room pressures, down to a near-vacuum of around one thousandth of an atmosphere.[6] These devices must be strong and well constructed, so avoid collapsing into the vacuum within.

The need for the application of heat in the process is still important in low-pressure diamond synthesis. This heat has a similar role as in high-pressure diamond synthesis, to encourage the carbon to re-align any bonds, and re-crystallize as diamond. Here the heat is applied to enable the hydrocarbons to decompose. The temperature varies, depending upon the pressure and hydrocarbon, and is typically between 600°C,[25] 700°C[35] to 800°C[1] and 1000°C.[6] These temperatures cause the molecules in the hydrocarbon to break down, and the result is a highly energetic all-carbon plasma.[2]

Process of deposition

Again, just as with high-pressure diamond synthesis, there is a 'formula' for success. Here we have a method, procedure and sequence that determine the end result. This includes the use of a catalyst. As the source carbon in this case is a hydrocarbon gas, a solid catalyst would not react in an appropriate way. Therefore a 'gaseous' catalyst is used in this gaseous environment.

The process is termed chemical vapour deposition and this accurately describes the nature of the reaction. The source hydrocarbon is carried in a steady stream into the apparatus and the carbon is induced to deposit on a given surface. To facilitate a simple description of the process and the apparatus simultaneously, Figure 14.1 is a schematic diagram of the CVD device. We will discuss the device, the role of each of

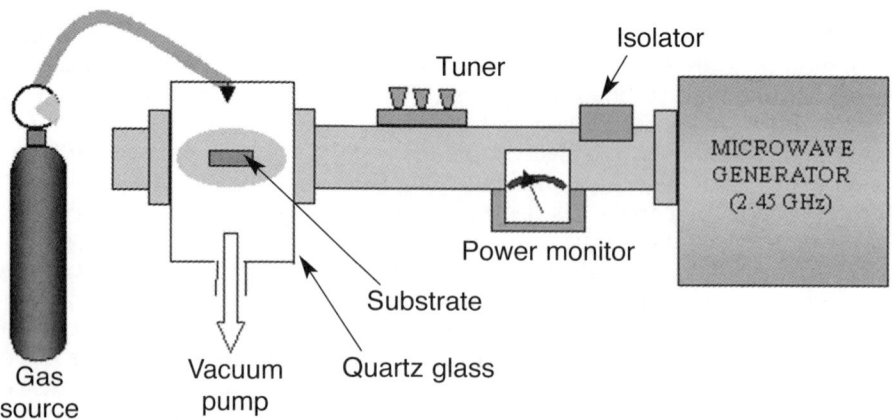

Figure 14.1 Schematic diagram for CVD (chemical vapour deposition) apparatus with a microwave generator energy source.

its basic components and the effect each has upon the process. This will lead us to a brief description of the chemistry involved.

This is a very simplified version of the system, but will serve for our purpose. The three important elements of the process are denoted by bold type. The **gas source** is the flow of hydrocarbon and catalyst gas, as we will discuss in more detail soon; the **vacuum pump** represents the induction of low pressure, and the **microwave generator** is responsible for the addition of heat.

The microwave generator produces heat by the same mechanism as that used in the microwave oven in our kitchens. This need not be the energy source, however, as other heaters are also applicable. Radio frequencies, electric heaters or tungsten filaments have also been used in CVD apparatus.[35] Oxyacetylene torches have also been used and have achieved good results.[25] The energy from the heating device is channelled, while under strict control (via isolator, tuner and power monitor), to the reaction chamber.[1] The heat produced in the generator reaches the reaction chamber where it comes in contact with the gases.

The hydrocarbon gases are fed into the reaction chamber by a pump, in combination with hydrogen. Early tests on the role of catalytic gases by Derjaguin proved that atomic hydrogen dissociated from the hydrocarbon was required, rather than molecular hydrogen.[93]

The stream of hydrogen carries the hydrocarbon into the reaction chamber and a vacuum then carries the remaining hydrogen out of the chamber following the decomposition of the hydrocarbon, thus continuing the process. The mixture is composed of approximately 0.1–1 per cent of hydrocarbon and 99.9–99 per cent atomic hydrogen. Hydrogen is believed to assist the process by helping to maintain the carbon atoms in the required tetrahedral arrangement. The result is a diamond lattice.[35] This is achieved by the hydrogen atom terminating any unresolved carbon bonds at the surface of the film. These carbon atoms are then unable to cross-link and form a graphite type bond.[94] The hydrogen also plays a role in etching away any graphite that may form leaving only diamond crystals,[25] and enhancing the quality and the growth rate of the diamond thin film.[2] Although it etches away both diamond and graphite, the growth rate of the diamond exceeds the etching process and therefore ensures that only a diamond film will form.[94] The microwave energy ionizes the gas into a carbon rich plasma of charged particles, hence CVD is also called plasma enhanced chemical vapour deposition (PECVD).[35]

The substrate is the name applied to the surface on which the diamond thin film is to be deposited. It may be composed of different materials including metals, and silicon based materials such as glasses.[94] The surfaces are roughened before deposition, usually by use of diamond powder.[1] The ability of this diamond thin film to adhere to different materials is the focus of most research. Currently the number of materials suitable for substrates is limited by these adhesion factors and the fact that the temperatures required for CVD are too high in many instances. They may melt during the deposition process. Some materials also react too strongly with the carbon, making them unsuitable.[94] The lowering of temperatures,[95–99] along with the effective isolation of the most productive and economically feasible pressure to be adopted within the reaction chamber, is the focus of much developmental research. These are among the issues facing the CVD diamond industry.[94,100–101]

In maintaining the low pressure of the reaction chamber, the vacuum pump serves to discharge unwanted hydrogen. The chamber may be formed from quartz plates as these provide a see-through alternative to the heavy metal jackets around earlier chambers. Visual technologies must be employed to enable viewing the process in

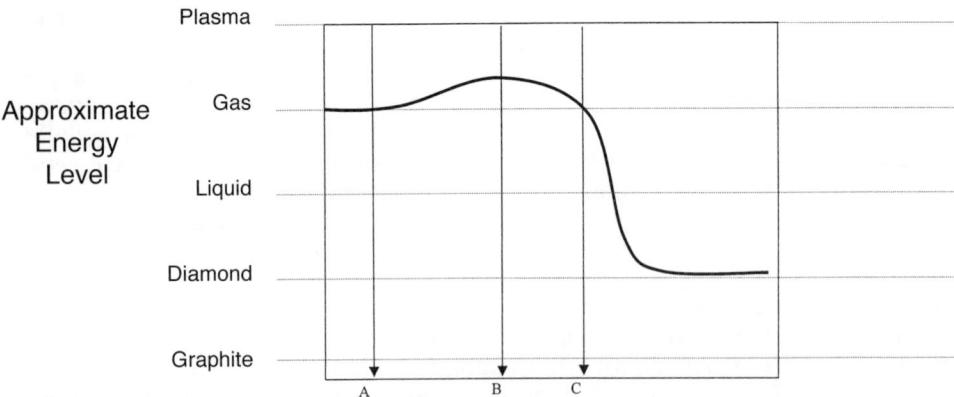

Figure 14.2 Conversion diagram for graphite to diamond conversion with starting conditions (A), activation conditions (B) and reaction conditions (C).

such devices. In reality the chamber looks nothing like the schematic in Figure 14.1 and is a very elaborate device.

Favouring diamond over graphite

Why does the carbon from these hydrocarbons deposit as diamond and not as graphite? Surely graphite is the thermodynamically favoured state at the low pressures of CVD techniques, just as it is if high pressures are used in synthetic diamond crystal development. Although graphite is the thermodynamically favoured state under these conditions, the secret rests in the fact that the carbon atoms contained within the hydrocarbons are already gases. They therefore already exist in a higher energy state than graphite, and indeed than diamond.

The reason for this apparently impossible conversion is that although graphite is the thermodynamically stable form of carbon in these conditions, the tetrahedral bonding of synthetic diamond is the kinetically favoured product.[25] The atoms arrange themselves in this way first. The activation energy required for the hydrocarbon to form the hexagonal lattice of graphite is higher, and the time required for this to take place is longer, than if the carbon were to form the tetrahedral lattice of diamond. Diamond is therefore the state of least energy. Remember the activation diagrams from Chapter 4? Figure 14.2 is the diagram for conversion from graphite to diamond as it appeared previously.

However, in the case of CVD this conversion diagram (Figure 14.3) looks very different. Remember that the starting carbonaceous material in this case is a hydrocarbon gas and not a solid such as crystalline graphite. Here the same notations are used as in Chapter 4. The energy is first applied at point A, exciting the atoms and elevating them into a higher energy level, approximating that of plasma, as the atoms are ionized by the heat. At point B the hydrocarbon decomposes and the carbon is deposited in the substrate as diamond at C.

The dotted line in Figure 14.4 represents the activation energy and the activation time required for graphite. Notice that the line is longer and the activation energy is higher than the level reached by the solid line corresponding to the formation of a synthetic diamond lattice.[25]

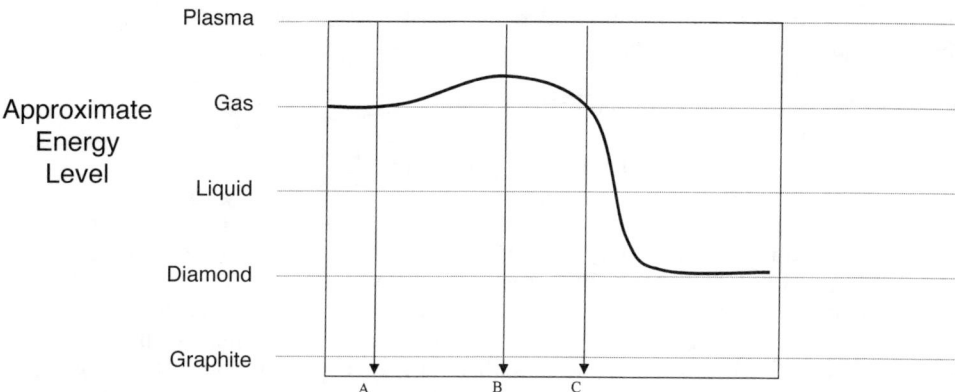

Figure 14.3 Conversion diagram for hydrocarbon to diamond conversion with starting conditions (A), activation conditions (B) and reaction conditions (C).

This rich plasma first nucleates on the substrate, then as the crystals grow larger, they coalesce to form a polycrystalline diamond layer.[1] The temperature must be slowly reduced to avoid the diamond decomposing to graphite. The result is a thin film of synthetic diamond on the substrate; however faults in the lattice are common, and perfect diamond is not always the final product.

Coatings and DLC

The deposited synthetic diamond from CVD processes can be controlled in a number of ways, all of which have an impact on the final product. The terms used to describe

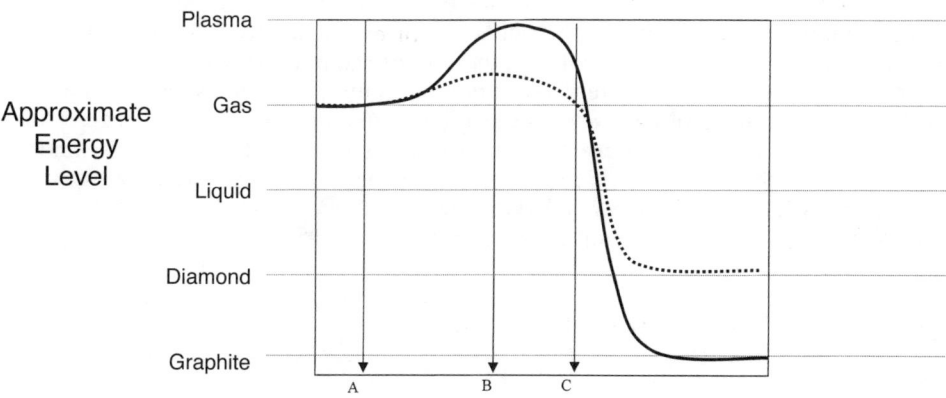

Figure 14.4 Conversion diagram for hydrocarbon to graphite conversion with starting conditions (A), activation conditions (B) and reaction conditions (C) (diamond conversion indicated with a dotted line for the purposes of comparison).

the result of the chemical vapour deposition of carbon on a variety of substrates depend upon a number of factors:

- the thickness and nature of the material;
- the purity of the material;
- electrical and mechanical properties of the material.

The thickness and nature of the material is in many ways related to electrical and mechanical properties. In this chapter we have used the term 'diamond thin films' which is only one of the possible products of CVD. Not only can the carbon be deposited to form solid crystalline films, but it can also be produced as ultra-thin coatings that are not removed from the substrate. Diamond thin films and diamond coatings are distinctly different products, with companies specializing in the production of one or the other. Both will be covered in Chapter 15 when we discuss the properties and gemmological observations of CVD diamond products and their applications.

The purity of the material relates to the percentage of actual synthetic diamond constituting the material and the quantity of other phases of carbon included. It is very expensive to ensure that only the tetrahedral lattice of synthetic diamond is allowed to form, without the infiltration of the hexagonal lattice of graphite or amorphous phases of carbon. The existence of these 'flaws' in the deposited synthetic diamond can result in imperfections that jeopardize the suitability of the product for a number of high technology applications.

These materials are broken into two distinct groups to represent the purity of the synthetic diamond in the film or coating. These are **diamond** and **diamond like carbon** or DLC. DLC is composed of a multitude of tiny diamond crystals all aligned together rather than one huge thin crystal. Additionally DLC may be included with metals (to create special properties such as semiconductors)[2] or other phases of carbon such as graphite, so it is not necessarily a pure diamond film.[35] The same technique can produce single crystals; however, they are extremely small and have few applications.

Large single crystal diamond thin films can be produced by other methods. These will be covered in Chapter 16 on alternative techniques of diamond synthesis.

The electrical and mechanical properties of the material are related to the end use of the diamond thin film or diamond coating, of either diamond or DLC. The temperature and the exact technique affects the size of the microscopic diamond crystals in the film or coating. Therefore these synthetic diamond products can be generated for exacting high-technology uses; they can be engineered. The average size of individual crystals is approximately 1.25 microns.[35] The size of the crystals, and the spaces between the crystals, are ideal for connections in electrical circuitry, especially if metal impurities have been included during growth.[25]

We continue to cover these topics as part of Chapter 15.

Chapter 15

Gemmology and diamond film technology

To follow on from our basic review of CVD processes and the production of diamond thin films and coatings, we will now look at the resulting products of this process. We will briefly delve into the properties of diamond thin films, from a gemmological perspective, and look at an industrial overview of the applications of diamond film technology. Once again it is important to note that this is a very complex and evolving science and the information contained here is by no means a complete account. References contained within the notes are suggested reading for those interested in pursuing this area further.

Applications of DLC and diamond thin films

Diamond CVD coated products are favoured due to their inherent hardness, durability and stable thermal and semi-conductive properties. Synthetic diamond coatings conduct heat five times better than copper, are harder than tungsten, and never wear out![25] The manufacture, development and designing of diamond coated products and the applications for diamond thin films is an ever-growing field. A listing of all the projects and possibilities for diamond film technology here could be out of date before it was read. In 1993 it was estimated that the CVD diamond market amounted to roughly US$30 million per annum, and was projected to reach US$1 billion by the year 2000. As the applications are so diverse and the industry so widespread, it is very difficult to gauge the current market worth and determine how close the global industry is to (or whether it has exceeded), this goal.

Obvious applications include mechanical parts such as bearings, seals and wear parts in the automotive, aerospace and naval industries. They are also used in weaponry. Wires for the electronic and telecommunications industry are also being diamond coated. These are being tested for use as hot filaments in heating mechanisms and potentially lamps.[94]

Diamond coated fibres and wires were produced by hot filament CVD method on a variety of core materials including tungsten and silicon carbide. This has been achieved with a diamond volume exceeding 95 per cent representing a high quality product. Possible future applications of diamond coated fibres include reinforcements in metal matrix composites (MMCs).[102]

However, wires and fibres are not the only industrial components benefiting from being coated with synthetic diamond. For example, Norton Diamond Film and Kennametal Inc. combined to develop diamond coated tools and wear parts such as seals. The aim was to produce parts that last 10 to 100 times longer than conventional parts. The three main focuses for the collaboration was to improve adhesion of the synthetic diamond to the metal substrates, to effectively coat round any other complicated three dimensional shapes, and to coat areas large enough to make the process economically feasible.[103]

Test-beds for this project included General Motors, Boeing and Ford, highlighting the potential for this type of technology for the automotive and aerospace industries. Other components and parts that are being coated with diamonds include silicon nitride ceramic ball bearings, produced by Wright laboratory under the sponsorship of Materials Directorate. These are said to last 100 times longer than standard steel ball bearings.

The computing and electronics industries are interested in the use of CVD in the production of computers and microchips. As such technological tools get smaller, the microchips are crammed closer together to increase speed and accuracy. This creates a great deal of unwanted heat that must be eliminated. A team of researchers at HiDEC have recently succeeded in securing another patent titled 'Method of Planarizing Polycrystalline Diamonds' that moves toward solving this problem. Because of diamond's excellent thermal conductive properties, the polycrystalline diamond acts as a heat conducting support in multi-chip modules (MCMs), allowing them to be made smaller while still retaining accuracy and speed for microprocessors in devices such as laptop computers and cellular phones. The team have perfected a way of growing the synthetic diamond thin film and using a polymer to fill the naturally occurring micro-cavities that cause service limitations.

Although much is known about CVD processes and diamond thin film deposition, there is still a long way to go. It is known that the crystallography of individual diamond crystals within the film can be controlled by modifying the heat and chemistry of the CVD technique. At low pressure and low temperature (slow growth conditions), the resulting film has triangular (111) faces, along with twin boundaries. If the percentage of the hydrocarbon is increased in the precursor gas mix, or the substrate temperature increased, (100) faces appear, both as square and rectangular forms.[94]

The current limitations to the expansion of the CVD diamond industry are the small variety of substrates that offer reliable and stable adhesion, and the rate of growth and quality of the resulting film or coating.[94] Benefits gained in one area of development often mean set backs in others. For example, to increase the quality of a diamond thin film may mean making modifications at the expense of the production time. Many novel and extraordinary methods are being trialed to maximize the potential of diamond thin films and coatings.

The rate of growth of the CVD process is defined as the total thickness of the film, or as the mass of the synthetic diamond deposited per unit time. It may therefore be possible to substantially increase the rate of growth by increasing the substrate area per unit volume. This has been the basis of development research projects, resulting in a new substrate being designed, with a three-dimensional pattern of wires and fibres.

Possibly the best part of this research was the claim that the increase in output was not made at the expense of gas flow or power consumption, thereby offering a more cost effective method of producing solid shapes and composites (MMCs).[104]

The technique of tiling is further example of innovation applied to solving another limitation in CVD diamond thin film technology. Tiling was a technique developed with the aim of increasing the size of diamond thin films while reducing the cost. The method makes use of the current diamond thin film production, utilizing special substrates, to produce diamond thin film with small surface areas. The tiling procedure

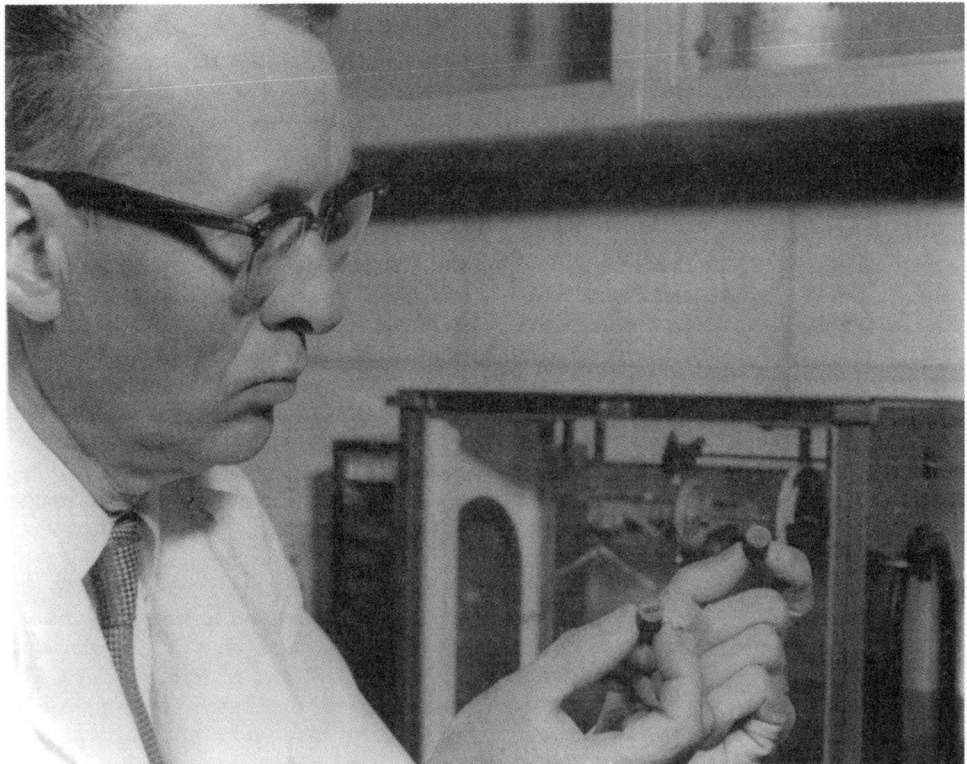

Figure 15.1 H. Tracy Hall at Megadiamond, holding two cutter inserts coated in polycrystalline diamond. (Courtesy H. Tracy Hall Foundation © 1998.)

involves growing these traditional diamond films, removing them from their respective substrates, and combining them. This is achieved by taking several films and arranging them in a mosaic design to form a new substrate. This new mosaic diamond substrate is then subjected to further CVD diamond growth to form one large diamond thin film. This process sounds revolutionary but studies are still underway to further improve the results, and to understand mis-orientation, strain, surface finish and crystallographic quality.[105–107]

CVD diamond thin film technology has advanced rapidly. In 1980 the reported growth rate of the average CVD diamond thin film was in the region of 0.1 µm per hour (or around 1 mm per year). There were problems distinguishing whether the resulting deposition was actually diamond, and nucleation was a source of many headaches in CVD laboratories.[6] Now CVD diamond thin film growth is making rapid progress. Growth rates are being reported in the range of 1000 µm per hour, but these films are deposited over small areas, and are of poor quality.[94]

Many universities and government departments have a vested interest in diamond thin film technology, and companies such as Applied Science & Technology Inc. (also called Astex) have nurtured a sub-industry selling CVD equipment to these research laboratories.[108] With many organizations manufacturing diamond thin films, what do diamond films look like, and what are their inherent properties in relation to other synthetic diamond products?

Visual characteristics of diamond thin films

Diamond thin films and synthetic diamond coatings are what the names suggest. The polycrystalline diamond layer applied to the substrate or to a specific item to be coated, can vary in thickness depending upon the end purpose. The thickness can be engineered to be anywhere between the approximate parameters of 10 nm[35] up to 1.5 mm.[109] Theoretically diamond thin films could be made to be very thick indeed; however, the cost would be enormous.

Visually the transparency and colour of a diamond thin film can also vary. A diamond thin film 1 micron (1 μm) in thickness was supplied to the GIA in 1989 by Crystallume of California. This film was found to be a light brown colour when viewed against a white background, in diffused light. This was attributed to an evenly distributed thin film interference phenomenon, of the type seen on optically coated camera lenses.[35] Diamond thin films with a pale grey appearance (to the naked eye), in diffused light, have also been observed by the author of this text. Additionally an extremely thin diamond coating applied to a transparent substrate was viewed by the author and was found also to be transparent and indistinguishable to the naked eye.

Colour in diamond thin film can also arise from impurities. In addition to the normal cause of colour in synthetic diamond (see Chapter 7), diamond thin films have impurities unavoidably included by another method. The substrate is formed of another material, usually silicon, and microscopic amounts of the substrate remain on the film when it is removed. This causes a hazy appearance and tell-tale coloration; but this is by no means a permanent problem. Many of the applications of diamond thin films require a perfectly smooth surface, but the films resulting from CVD processes are far from smooth; the films must therefore be polished. The process of polishing a diamond thin film removes the fine layer of diamond film that is contaminated with the residual substrate. Both surfaces must be polished to completely eliminate the haziness and coloration.[35]

Under normal light a different phenomenon can be seen in some diamond thin films. Interference colours can also be observed in the Crystallume sample thin film across the entire sample. This rainbow of interference is due to the fact that the sample thickness of 1 μm is in the same order of magnitude as the wavelengths of visible light.[35] These interference colours are not seen in diamond films 3 μm to 4 μm in thickness.[35]

Other visual characteristics of diamond thin films and diamond coating will be discussed in context in the section 'Diamond thin films and coatings in the gem industry'.

Gemmology of diamond films

The polariscope provides not only a standard gemmological test to determine birefringence and axiality, but also an important picture of the internal structure of the material under test. When viewed under a polarized light source, the film analysed by the GIA was found to exhibit no extinction, as would be normally expected for an isometric gem material such as diamond. However, diamond thin films and diamond coatings are polycrystalline, and due to their aggregate composition of randomly oriented diamond crystals, the result is in keeping with the 'all light' result observed in other polycrystalline substances. In the case of the diamond thin film, the aggregate reaction is due to light scattering at the grain boundaries, with the possible addition of phenomenon as a result of strain.[35]

In contrast to their mono-crystal synthetic diamond cousins, diamond thin films are not easy to test gemmologically. The fact that they are inherently adhered to their respective substrates (unless removed chemically), and that the coatings are permanently attached, means that testing as mentioned above becomes near to impossible.

The substrate or 'base' material will also show a result in polarized light (and potentially in other tests), which could be confused with a reaction of the film. Such a 'false' result could adversely affect tests intended to identify the presence of such a film.

Tests that do not rely on the visual or optical properties of the diamond thin film would be a better choice in this situation. We must then turn our attention to the thermal and electrical properties of diamond, and therefore diamond thin films. As we know from previous chapters, diamond has extremely high thermal conductivity, but no electrical conductivity, unless engineered as (or occurring naturally as) type IIb diamonds which are electrical semi-conductors. The most commonly applied gemmological test to check for thermal conduction is the diamond tester, or diamond probe. The GIA used such an instrument in their tests on the diamond thin film supplied to them and recorded the following results.

Several spots on the diamond thin film were tested to ensure that any results were not a localized phenomenon, but were representative of the diamond thin film as a whole. The researchers used the GIA-GEM Duotester and results were ambiguous. The tester showed readings of between 60 and 80 on the scale, which corresponds to the lower end of the green zone representing 'diamond'. This was a weak result, as diamonds usually give a result in the vicinity of 80–110 on the tester scale. The problem arose because the silicon substrate gave exactly the same results as the diamond thin film![35]

It is a known fact that diamond thin films and diamond coatings conduct heat.[110] This is one of the properties that form the basis for their use in industrial applications.

The deduction was made that this diamond thin film was too thin to achieve a clear result, on this particular type of instrument. The results were probably due to the substrate and not the diamond thin film. As diamond has a higher thermal conductivity than the silicon used as the substrate, the conclusion was reached that the thickness of the film (1 µm) was not sufficient to disguise the effect of the substrate.[35] The problem is caused by restricted heat flow to the thin film as heat flowing to the film is unable to dissipate into the substrate, so the film heats up the probe and is effectively measuring the thermal conductivity of the substrate. Therefore very thin films of this type would need to be tested on a heat sink such as the aluminium plates supplied with many diamond probes for best results. This has ramifications for the potentially pernicious application of diamond thin films, discussed in the final section of this chapter.

While testing for electrical conductivity in diamonds is not a test commonly performed in gemmological laboratories, electrical activity in minerals and synthetic materials is an important property that is perhaps more highly appreciated by scientists. As we know that diamond is an electrical insulator, it is of interest to us to determine whether diamond thin films exhibit the same insulating properties.

When connected to two steel electrodes, the 1 µm thick diamond thin film on a silicon substrate, analysed by the GIA, showed a weak electrical conduction in some places. The results corresponded to a reading of 1 to 2, while placing the two electrodes together to give the highest possible result recorded a reading of 130 on the same scale. As we have discussed briefly in reference to DLC, pure diamond thin films are hard to achieve. The weak electrical conductivity of the diamond thin film tested by the GIA may be due to small amounts of 'a more graphitic phase' of carbon at the boundaries of the diamond grains.[35] Graphite is an electrical conductor, as well as a thermal conductor.

Other standard gemmological tests have been applied to diamond thin films and their constants evaluated. In 1991 GIA published in the journal *Gems and Gemology* the refractive index for a DLC thin film as being 2.00 which is over the range of the standard refractometer. They were also described as being amorphous, brown in colour and with a hardness between 9 and 10 on the Mohs scale. Earlier evaluation of diamond thin films (rather than DLC) found the refractive index to be higher, at 2.4, crystalline, bluish-grey in colour, and with a hardness of 10. These differences in refractive index and hardness are probably due to the purity of the film, the residual amount of hydrogen, and graphite component in the DLC films.[37]

Diamond thin films and coatings in the gem industry

Although this diamond thin film technology is challenging, and the direct applications for industry seemingly endless, one may well ask how this may fit into the gem industry.

Diamond films and DLC do not seem to fill many of the criteria to be classed as a gem material as they are not particularly beautiful, nor are they rare. They are however suitably durable.

As the average gemmologist, lapidary or rock hound knows, there are many beautiful gemstones that are simply not practical for wearing. Indeed there are many gemstones that are worn as jewellery that are soft and brittle and must be treated with extreme care. Natural emeralds and tanzanite spring to mind. Imagine the range of beautiful and glamorous jewels that would be available if they all had the durability of diamond. Imagine if these fragile gems could be coated with diamond to protect them and to preserve their beauty. Imagine the possibility of coating opal to 'seal in its water' and prevent crazing.[35]

This dream may soon become a reality. Currently patents already exist in Germany to coat such soft and beautiful gemstones, for example apophyllite and kyanite; however, no products are known to be currently available.[111] DLC films have been applied to citrine, amethyst, beryl, tourmaline and garnet. These coatings had a thickness of 0.08 μm.[37] Note that these are silicates. Studies have also been done on achieving diamond coated sapphires. Problems exist at the interfacial layer with the adhesion of the diamond to the sapphire, and some surface cracking. This is the main problem confronting research. Diamond films adhere to some substrates while other substrates, such as sapphire, prove problematic.[112] Additionally as the gems must be heated to around 1000°C to induce deposition, this limits the procedure to those gem materials able to withstand such temperatures.[37]

Once these problems are overcome, not only will industry be able to produce its ultimate materials, but jewellers may finally have their ultimate gems. Has the door been opened for a whole new era in gemmology? What may the future hold? Synthetic diamond coated simulant?

Diamond coated cubic zirconia is believed to be improbable[35] as cubic zirconia is composed of zirconium oxide which has a vastly different atomic structure, and, to date, oxides have proved problematic as substrates. However, one can never be too careful. A diamond coated diamond simulant, thought to be cubic zirconia, has been reported by a member of the jewellery industry.[113]

The diamond simulant strontium titanate, with its high dispersion and faint yellow tinge, has been successfully coated with synthetic diamond. Again the thickness of these films is around 0.08 µm. It is very important to remember this as many companies in the gem industry handle large parcels of melée, which could easily contain synthetic diamond coated strontium titanate! Stones as small as 0.002 ct have been reported as being of this type.[37]

This leads us to the next question. How do we detect treatments in gemstones if they are coated with an impermeable layer of synthetic diamond? Certainly this proposition could have benefits, as many of the unstable treatments may now become stable and permanent. This does not necessarily make these treatments acceptable.

Treatments such as the oiling of emeralds and heat treatment of corundum have been generally accepted. However those not generally accepted in this way will still require identification and full disclosure. Impregnation of resins and fracture filing of diamonds could possibly take on a whole new face in the future.

To date the most commonly coated gemstones are diamonds themselves. Many companies have conducted testing into synthetic diamond coating on natural diamonds. Why apply a synthetic diamond coating to a gem that is already diamond? To improve the colour of a faceted stone, or to increase its weight. Currently technology has not supplied a method that offers diamond thin film of high enough quality, at a cost reasonable enough to make coating faceted diamonds to increase weight economically feasible. However in the future the possibility of increasing the weight of a faceted diamond from 0.99 ct to 1.00 ct is not unrealistic.[35]

Coating diamonds to improve their colour is a procedure that has been practised for many years. Yellowish coloured natural cape diamonds have been coated with a number of substances to improve the colour of the stone. Most of these treatments have proved to be unsuited for the task, and are obvious to the observer. The most convincing coating used over the years is the metallic fluoride[35] used to coat transistors, resulting in a coating similar to the blooming on camera lenses.[37] The colour of the natural diamond is improved because the coating is blue in colour, and blue is the complementary colour of yellow in the spectrum. The blue colour therefore has the effect of 'cancelling out' some of the yellow tinge in the stone.

Synthetic diamond coatings are potentially the new generation of colour enhancing coatings. Just as synthetic diamond mono-crystals can be doped with boron to make them blue in colour, so too can synthetic diamond thin films. The boron is added in the form of boron and hydrogen gas B_2H_6 and is mixed into the precursor gases. The result is a bluish semi-conductive synthetic diamond thin film or coating.[94] In 1989 Sumitomo Electric reported the successful coating of a near colourless natural diamond crystal with a blue synthetic diamond coating as thick as 20 µm. The result was a blue coloured electrically conductive diamond crystal that looked relatively convincing.[35]

The first reported observation of such coating on a faceted gem diamond appears to be in 1984, when the GIA evaluated a synthetic diamond coated natural diamond that was believed to be approximately H to I in colour before coated, but subsequently appears as a G colour.[37] This was again the case with stones submitted to the GIA in 1991.[37]

Diamond coatings applied to improve colour, or as a mask on other gems, or of diamond simulants is not as alarming as it may at first appear. There are methods to detect such a treatment! First, as was the case with the faceted diamonds coated with blue synthetic diamond thin films observed by the GIA, the stones appeared to have a suspicious grey tinge to them, while others exhibited an obvious dark bluish-grey coloration.[37] This is the first hint of diamond coating.

Second, the colour in natural blue diamond is not entirely homogenous, but has a slightly patchy appearance. Bluish diamonds coated with boron doped synthetic diamond will exhibit the shaped edges of the coating.[35] Look also at the facet edges and junctions for evidence of fluctuations in the distribution of the coating on the stone surface. This can be made easier by immersing the stone in methylene iodide.

Third, the diamond probe is always the best defence in detecting diamond coatings on simulants and other gemstones. Diamond coatings of the type discussed here have an average thickness of under 2 μm, and, as mentioned in the previous section, do not test as diamond when probed with the diamond tester. It has been calculated that coatings would need to be a minimum of 5 μm to register as diamond with this instrument.[37] While coatings of this thickness are not impossible, they are unlikely, as these films are more expensive to produce at this time.

To summarize, the GIA have said that they believe the current synthetic diamond thin films should not present problems as detection is straightforward and easily achieved with standard methods.[35]

A new gem for the future?

While diamond coatings may have a number of potential applications in the gem industry, until recently the thicker diamond thin films have been seen as purely a commercial and industrial tool. That was until the synthetic CVD diamond expert, Dr Bachmann from the Phillips Research Laboratory at Aachen in Germany, quite literally cooked up a present for his wife.

Dr Bachmann commissioned a set of two-tone jewellery set with plates of synthetic diamond thin film. The square pieces were cut by lasers in two sizes, of a final weight of 1.00 ct and 0.21 ct respectively. The original diamond thin film was 40 mm in diameter, 0.25 mm in thickness and weighed 5.5 ct in total. The diamond film was of a grey drusy appearance, and is composed of polycrystalline diamond and light is scattered from the surfaces of each microscopic crystal. It is said that when polished such diamond thin films become as transparent as glass of a similar thickness.[114]

These particular films were grown by a CVD process of the type we have already discussed. The precursor gas was composed of 2.8 per cent methane in 98.2 per cent atomic hydrogen at a gas pressure of 180 mbar. The substrate was heated by a microwave induced plasma to approximately 900°C. Raman spectroscopy has determined that the tiny synthetic diamond crystals within the film were very well formed with excellent crystallinity. Thermal conductivity tests obtained results of 2200 W/m°C, comparable with natural type I single crystal diamonds.[115]

This diamond thin film may sound large at 5.5 cts, but it is by no means the largest produced. In 1995 researchers at the University of Florida, J. Adair and R. Singh, patented the process for the production of what was considered at the time to be the world's largest synthetic diamond. Grown by a CVD process, the 1600 ct diamond was a polycrystalline plate measuring 11 inches in diameter and 1.5 mm in thickness.[109]

CVD and high-pressure synthesis are not the only methods of diamond synthesis. In Chapter 16 we will briefly look at a few notable methods of making synthetic diamond.

Chapter 16

Other methods of diamond synthesis

We have covered two major methods of making synthetic diamonds. These were high pressure and low pressure. This can be subdivided into more exacting methods. The high-pressure methods include the belt apparatus, the split sphere or BARS devices and the tetrahedral press. The low-pressure methods include the microwave and radio frequency (RF), the tungsten filaments, and the plasma torch heating systems.

In addition to the use of plasma torches, oxyacetylene torches are also widely used in the synthesis of diamond thin films. This method is relatively popular because it offers low setup costs, ease of control and exceptionally high growth rate of the film.

The experimental apparatus consists of a commercial oxyacetylene torch connected to a stage, a set of mass flow controllers for O_2 and C_2H_2, a T-shaped copper block (cooled by water) and a two-colour pyrometer to measure the substrate surface temperature. The mass flow controllers are under computer control to cycle the flow ration of the precursor gases O_2 and C_2H_2. The growth of the diamond thin film is by achieved the cycling of the flame.[116,117] The result is a mosaic structure, with the unusual occurrence of a structure known as a 'diamond bud'.

There are, however, a number of other methods of producing of synthetic diamond. University and private projects have experimented with new methods with varying success. We will briefly describe here a few of the more commercially viable or academically acclaimed, or interesting methods.

Synthetic diamond plates and laminates

The aim of this method is the production of diamond windows suitable for visible and infrared optical devices. It was patented on 22 June 1993, by the National Institute of Standards and Technology in the USA. This invention overcomes the problem of surface roughness that results from other methods of forming diamond thin films from a gas phase.[118] Refer back to the figures contained within previous chapters. Diamond is a special substance that is transparent to many forms of electromagnetic radiation such as high-energy gamma rays, X-rays, ultraviolet light, visible light, infrared radiation, radio frequency and microwave radiation.[119] The techniques covered by this invention allow for the production of diamonds, plates and laminates which have two smooth parallel surfaces, perfect for the transmission of visible and infrared light. It is hoped that this invention will allow for simpler manufacture of optically transparent diamond windows with a thickness up to several millimetres or greater.[118]

The process involves placing two planar substrates parallel to each other with a distance between them equal to the desired thickness of the film. A diamond thin film is then deposited on both substrates simultaneously by standard CVD methods. The deposition is continued until the two diamond thin films merge into a single diamond plate.

The patent also describes a method of preferentially heating selected areas of the substrate. This provides for uniform growth and prevents the formation of voids in the finished film.

The invention eliminates the need for extensive post-formation processing such as polishing. Polishing is required in the case of regular synthetic diamond thin films, to provide smooth surfaces. Thus this method produces optically transparent diamond plates at lower cost than earlier methods which required polishing of at least one surface.[118] The diamond plates produced by this method are suitable for any purpose requiring hard, resistant, optically transparent windows.

Shock-wave process

When one thinks of shock waves it is hard to lose the mental image of the violent explosions used in mining or demolition. This is especially the case when associating diamonds with explosive forces. However, for an explosion to be powerful does not necessarily mean it has to be large and dramatic. The shock wave process used in the synthesis of diamonds is small, contained completely within a laboratory, yet the results are a tangible and glistening example of their awe-inspiring power.

In 1961, P.S. De Carli and J.C. Jameson were the first to successfully use high explosive shock waves to synthesize diamonds. The explosion created during their initial experiment lasted one millionth of a second and applied 300 000 atmospheres of pressure to the graphite source material. The shock wave process has improved with advanced technology.[5] Commercial processes of this type are undertaken by Du Pont and Allied Chemicals at temperatures of 1000°C and explosive pressures of 280 000 to 480 000 atmospheres within a hollow tube of explosive containing graphite impregnated copper. The process takes one-fifteenth of a second, and the slowly cooling copper avoids the major problem of the synthetic diamond reverting

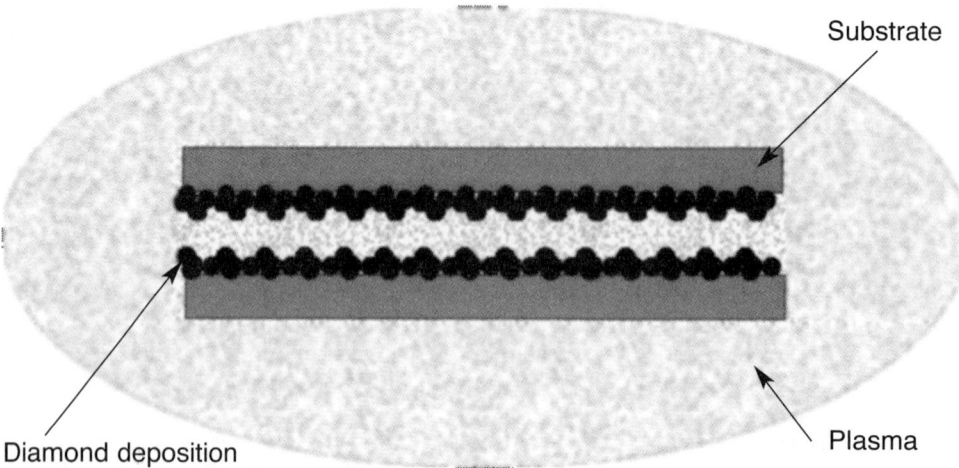

Figure 16.1 Schematic representation of the dual deposition involved in producing diamond film laminates.

to graphite. The resulting microscopic particles, suitable for use as abrasives, apparently contain synthetic diamonds (isometric tetrahedral lattice) and lonsdalite (hexagonally bonded diamond).[5, 120–124]

Ion implantation

This technique was developed by J.H. Freeman at Harwell Atomic Energy Research centre supported by the National Research Development Corporation in the United Kingdom in 1977. It involves the production of a beam of carbon ions (charged atoms) in a vacuum chamber, and the acceleration of this beam by an electrical voltage. The arc shaped chamber is typically small enough to fit comfortably on a large dining table, with additional controlling instrumentation. As this method is performed in a vacuum, this is truly 'low-pressure' diamond synthesis.

Initially the stream is carried with other precursor gases through the chamber. To assist in purification of the carbon film, the stream is passed through a magnetic mass-spectrometer, which separates the carbon ions from all the other accompanying particles in the gas. The remainder of the carbon stream is used to bombard a substrate of seed. The energy in the carbon ions is high enough to cause compressive stress, and thus hydrostatic pressure.[125] This causes them to be *implanted* into a diamond seed crystal or substrate surface. This is very different from the surface properties of the more common CVD method, in which the carbon is adhered as a layer onto the surface of the substrate, rather than implanted into it.

At room temperature this merely generates amorphous ion implantation, a feature that alone is a very useful phenomenon. But if the procedure is conducted under temperatures of 700°C, the gas stream becomes a plasma stream, and the ions (subject to the hydrostatic pressure conditions at the interior of the film) arrange themselves into the tetrahedral covalent bonds of diamond.[125] The result is a synthetic diamond layer *imbedded* into the outer surface of the substrate. All problems associated with adhesion are eliminated in one move. It is not fully understood exactly what occurs at the interfacial layer; however, this gives a rough guide. These ion imbedded diamond thin films are very thin indeed, thinner than average CVD films. However, they are much higher in purity with a higher percentage of diamond.[126, 127]

This has excellent applications in the electronics and medical industries. For example diamond films implanted into surgical implants such as replacement joints, are very efficient at ensuring that the patient's bone does not graft itself to the implant (thus seizing up the new joint). To achieve the opposite result, with the bone grafting to the implant, cobalt is used in place of diamond.

Scottish doctors have used synthetic diamond coatings in replacing hip joints, to extend their wear-life; using this technique they should last years longer than present replacements. Dr P. John of Heriot-Watt University said the synthetic diamond was used to cover the ball in the ball-and-socket hip joint replacements, providing protection and reducing wear. Hip replacements are one of the most common surgical operations in the western world because the natural joint is easily worn out or is affected by arthritis.[128] Perhaps this is a problem of the past!

Single crystal diamond thin films

The first single crystal synthetic diamond coating and films were grown in 1991.[25] This method is similar in many ways to the previously mentioned low-pressure ion implantation. Prior to the development of this technique most diamond thin films were a mosaic of polycrystalline diamond particles, randomly oriented, and together forming the diamond layer.

In this technique an accelerator is used to insert a stream of carbon ions into a copper surface using a method similar as described below. The ions form a density of about one billion (one thousand million) ions per square centimetre, embedded into the surface of the copper substrate. The carbon-laden copper is then blasted with a series of powerful laser bursts, which will rapidly liquefy the top-most layer of the copper. Once the laser pulses stop, the copper slowly solidifies, and the carbon ions migrate to the surface of the copper. Here they align themselves in the tetrahedral bonds of diamond, thus forming a solid, although very thin, synthetic diamond film.

As they all align simultaneously orientation is uniform, resulting in the formation of a single crystal.[25] This method is still in its infancy and applications are under investigation.

Diamonds and buckyballs

We briefly mentioned buckminsterfullerene in Chapter 3, and hinted at the use of these fullerenes in the synthesis of diamond. A group at the Centre Des Recherches sur les Très Basses Temperatures, in Grenoble, headed by Manuel Núnez Regueiro, have successfully pressurized buckminsterfullerene in a diamond anvil,[129] and made a remarkable discovery. They found that it suddenly transformed to diamonds, an occurrence of potential interest to a variety of industries. Scientists already knew it was possible to transform graphite in this way (within a diamond anvil), but the buckminsterfullerene transition took place at lower pressures. Additionally this transformation took place without any form of catalyst or precursor gases.[130] Indeed this was the type of 'magical' transformation dreamed of by early alchemists, and scientists of the late nineteenth century during their quest for diamonds. Experiments have commenced to analyse the metamorphosing of C_{60} into diamond, with a number of papers emanating from this research.[131]

Other fullerenes, especially C_{70} (a 70-atom spherical carbon molecule with similar structure to buckminsterfullerene), have also been found to have interesting features for diamond synthesis. Scientists at Northwestern University in Illinois found that these fullerenes assist the growth of diamond thin films, which may be used to harden the edges of tools or to create electrically insulating barriers in electronic circuitry.[132]

Fullerite is full of diamonds

Following the discovery that buckminsterfullerene alters to diamond under high pressure, researchers began to investigate the effects of similar pressure on fullerite. Fullerite is the solid form of buckminsterfullerene.[4,133,134] In September 1995, the United States Patent Office granted a patent for a method of producing diamond crystals from metallofullerite matrix, amazingly without the need for high-pressure apparatus.

The method consists of heating a portion of metallic solid, comprised of a metal carbon matrix of an allotropic metal and metallofullerites of that allotropic metal, to (or near to) the critical temperature for its percentage of composition of the allotropic metal, carbon, and other effective ingredients. After the portion has been heated, hydroquenching is undertaken. The quenching is conducted rapidly enough to collapse fullerene structures present in the matrix into diamond crystals. The resulting metal-carbon matrix of an allotropic metal, metallofullerites of the allotropic metal, and diamond crystals are also disclosed.[135]

It is interesting to note the similarities between this modern method and the work of Moissan, with his heated rod of metal and carbon, quenched to cool them rapidly (see Chapter 1).

DIY diamonds?

Even though technology has come a long way, synthetic diamond research and production is still expensive. Equipment is out of the reach of many laboratories (let alone the average man or woman!). Imagine if this barrier were removed, and synthetic diamond could be created from the humblest of substances.

In 1993 a team at Pennsylvania State University lead by Dr P.A. Bianconi invented a novel method for arranging the carbon atoms in a polymer that mimics the carbon structure of crystalline diamond. When this polymer is cooked at a relatively low temperature, it is largely converted to diamond! This polymer is manufactured from plastics and the temperatures equate to those of a home oven!

Following successful laboratory synthesis of tetrahedral silicon and germanium polymers, Dr Bianconi conceived the idea that analogous polymers might be made out of carbon. The result would be a tetrahedrally bonded molecule of carbon atoms, in a polymer string. She theorized that by roasting away all the atoms in the polymer, except those of its skeletal tetrahedral carbon structure, the substance remaining should be tetrahedral carbon, in fact diamond!

The recipe for these 'home cooked' diamonds begins with a commercially manufactured compound called alpha-alpha-alpha-trichlorotoluene, whose molecules can easily be made to hook together as polymers. Ordinarily, the polymers they form are polyacetylene-compounds in which the carbon atoms are linked together as interconnected chains, which, when roasted, yield only humble graphite. However, the Penn State chemists prepared a violently reactive alloy (or reducing agent) from two soft metals, sodium and potassium, and mixed them with the trichlorotoluene starting material. This concoction was homogenized using an intense bean of ultrasound. It was this that succeeded in synthesizing the desired polymer.

The compound is a kind of proto-diamond, in which the carbon atoms are bonded in interlocking tetrahedrons, among neighbouring atoms. Low temperature heating of this polymer drives away most of the unwanted atoms, leaving a transparent, ultra-hard residue – diamond.[136] These temperatures were as low as those obtainable by the common kitchen oven. In the words of Dr Bianconi, 'What we have made is poor quality diamond, a mixture of true diamond with diamond-like carbon and ordinary carbon … Our diamond is not as good as the synthetic diamond films others have learned to make during the last 10 years by depositing carbon vapor on a variety of surfaces. But we merely wanted to show that it is possible to make diamond from a synthetic polymer, and we used the crudest possible techniques to demonstrate the principle. We know they can be refined to yield far better results.'

Dr Bianconi said in an interview that the Penn State group had withheld its announcement for 18 months to test and reconfirm its findings. The group has applied for a patent.[137]

Epilogue

Can we forecast the future for synthetic diamonds? And what will be their impact on the jewellery industry? One can only speculate. Many of the major barriers that seemed scientifically insurmountable in the past, are now mile-stones along the road to success.

High-pressure researchers who once dreamed of making diamonds have now moved on, with the search for metallic hydrogen among the new goals. Diamond synthesis is still a thriving industry among academic and commercial researchers. While in the past purely scientific research has, to an extent, driven commence and industry with respect to new developments in super-hard materials, more recent times have seen the reverse. More and more frequently it is industrial needs for better materials that are setting the pace of research, and although there is still much to be learned and refined, synthesis of diamonds is now a commercial industry in its own right. However, new research undertaken by the Carnegie Institute in Washington has found that diamonds will 'bend' at high pressures. R. Hemley, D. Mao and G. Shen, collaborating with scientists at the Ecole Normale Supérieure du Lyon in France, and the European Synchrotron Radiation Facility in Grenoble, have devised a technique for predicting the distortion of the diamond lattice at high pressure. At around 200 to 300 gigapascals (approximately 2 to 3 million atmospheres) the strongest material will bend, but not break![139] Could diamond lose its throne as the king of minerals in the future?

Costs have been reduced. Once costly high-tech instruments are now available for sale to laboratories and factories around the world. Indeed, as we have seen, new research means that costly instruments and devices are not necessarily needed at all! Diamonds made without pressure,[139] or cooked in the home oven.[137] While remarkable, are discoveries opening the door to an uncertain future for the gem industry? Could a gem be wrapped in a polymer and cooked in the home oven to coat it in synthetic diamond?

While most of us are not in a position to make any kind of prediction, one thing is certain. Gem quality near colourless synthetic diamond crystals and faceted stones are a fact of life. They are our reality. Our insurance is based on the knowledge we have built from observation. These stones yield certain characteristic properties and features that can be identified. With the information contained within this text, the stage is set for a practical approach to be adopted when confronted with the task of differentiating a natural diamond from a synthetic one.

We have the means and the technology; however, the most vital ingredient is diligence. This includes continually up-dating our knowledge base and keeping abreast of vital developments. While our current synopsis may seem all encompassing, new discoveries come to hand all the time. These may also offer the prospect of applications to the gem industry, and a whole new realm of unknown quantities for the future. For example, while we know that synthetic diamonds contain metallic inclusions (derived from the metallic catalyst), and natural diamonds do not, studies of synthetic diamonds in 1995 revealed silicate phases of inclusions.[140] Silicates! Aren't they meant to indicate a natural origin? These inclusion phases, while small, were definitely present. This is just the beginning. Synthetic diamonds have been successfully grown not from a metallic catalyst, but from a volatile-rich kimberlite melt.

A series of experiments were carried out in 1993 using high pressure and high temperatures, applied with a modified belt type apparatus. The source material was a sample of an aphanitic group 1 kimberlite from the Wesselton Mine in South Africa. It was found that the kimberlite had a strong solvent/catalytic effect in the synthetic diamond formation, giving vital information about the origins of natural diamonds. However, the resulting synthetic diamond had well-developed octahedral faces {111}, and its overall morphology resembled that of a natural diamond. The morphology of the synthetic stone differed greatly from that of synthetic diamonds grown with the usual metallic catalysts.[141] Could this mean the introduction of synthetic diamonds that are not disguised, but rather may intrinsically adopt the appearance of natural stones?

Only time will tell whether these new developments will have a direct impact on the gem diamond industry. The next millennium will no doubt hold new and promising applications, products and improvements in the field of diamond synthesis. Theories supported by many scientists suggest that diamond may already have lost its place on the 'hardest material' throne. Perhaps the synthesis of carbon nitride will become the new industrial favourite in the future. Indications suggest that it represents a 'harder than diamond' alternative to currently available products.[142,143]

Synthetic diamonds and synthetic diamond thin films may mature to become a friend or foe to the gem diamond industry. Experience gained with the introduction of other synthetic gems and enhancement treatments (fracture filled diamonds for example), show that there is a very fine line between a devil and a darling. Inventions that offer infinite promise and potential are often the bearers of infinite problems in our professional lives.

Already synthetic diamonds have infiltrated the diamond market under the guise of natural diamonds. The first instance of this occurring with rough synthetic diamond was reported in 1995. Here micro-crystals were purchased in Canada, and were said to be from the core drilling in Saskatchewan. The fraud was detected when distinctive habit and metallic inclusions were observed in samples.[37]

I trust that this compilation can serve as a tool to gemmologists, students and those in the gem and jewellery industries, and may shed some light on what can be done to control our destinies in a future shared with synthetic diamonds.

Notes

1 Collins, A.T. (1988) Diamonds in Modern Technology: Synthesis and Applications. In *The Nature of Diamonds* (G.E. Harlow, ed.), Cambridge University Press.
2 Hazen, R.M. (1993) *The New Alchemists: Breaking Through the Barriers of High Pressure*. Random House, Time Books.
3 Mellor, J.W. (1943) *Mellor's Modern Inorganic Chemistry* (rev. by G.D. Parkes), Longmans, Green & Co.
4 Aldersey-Williams, H. (1995) *The Most Beautiful Molecule: The Discovery of The Buckyball*. John Wiley & Sons Inc.
5 Webster, R. (1983) *Gems: Their Sources, Description, and Identification* (4th edition, rev. by B.W. Anderson), Butterworth & Co Ltd.
6 Nassau, K. (1980) *Gems Made by Man*. Gemological Institute of America.
7 Read, H.H. (1946) *Rutley's Elements of Minerology*. (23rd edition), Thomas Murby and Co.
8 (1993–96) 'Carbon Cycle. *Microsoft® Encarta® 97 Encyclopedia*. Microsoft Corporation.
9 Gribbin, J. (1995) *In Search of the Double Helix: Quantum Physics and Life*. Penguin Books.
10 Kirkley, M.B., Gurney, J.J. and Levinson, A.A. (1991) Age, Origin, and Emplacement of Diamonds: Scientific Advances in the Last Decade. *Gems and Gemology*, **27,** 1, 13.
11 Mykolajewycz, R., Kalnajas, J. and Smakula, A. (1973) *Journal of Applied Physics*, **35**, 1773.
12 Knoop, F., Peters, G.G. and Emerson, W.B. (1939) *Journal Res. Nat. Bur. Standards*, 23.
13 Schumann, W. (1977) *Gemstones of The World* (English edition trans. E. Stern), Sterling Publishing.
14 Clark, C.D. and Berman, R. (1965) *Physical Properties of Diamond*, Clarendon Press.
15 Winter, M. (1996) Web Elements. Department of Chemistry at the University of Sheffield.
16 (1993–96) 'Pressure'. *Microsoft® Encarta® 97 Encyclopedia*. Microsoft Corporation.
17 Morrison, P. (1996) Wonders Under Pressure. *Scientific American*, **6**.
18 Scandolo, S., Chiarotti, G.L. and Tosatti, E. SC4: a New Form of Carbon Metastable at Tpa Pressures. International School for Advanced Studies (SISSA), Italy.

19 Teter, D.M. A New Post-B Diamond Carbon Structure: Lessons from Silica. Geophysical Laboratory and Center for High-pressure Research, Carnegie Institute.
20 Mitchell, J.W. (1983) *Energy Engineering*. Academic Press.
21 Quinn, T.J. (1990) *Temperature*, 2nd edition. Wiley.
22 Hatsopoulos, G.N. and Keenan, J.H. (1981) *Principles of General Thermodynamics*. Wiley.
23 Van Wylen, G.J., Sonntag, R.E. and Borgnakke, C. (1994) *Fundamentals of Classical Thermodynamics*, 4th edition. Wiley.
24 Harlow, G.E. (1998) What is Diamond? In *The Nature of Diamonds*, Cambridge University Press.
25 University of Wisconsin (1998) Buckyballs, Diamond and Graphite. www.chem.wisc.edu/~newtrad/CurrRef/BDGTopic/BDGtext/dmndref.html
26 (1993–96) 'States of Matter'. *Microsoft® Encarta® 97 Encyclopedia*. Microsoft Corporation.
27 Bundy F. (1960) Patent No. 2947611 to General Electric, Diamond Growth using platinum at higher temperatures, US Patent Office.
28 Wentorf, R. (ed) (1962) *Modern Very High-pressure Research*, Butterworths.
29 *Indiaqua* (1990) De Beers Public Relations Magazine.
30 Banholzer, W. (1991) *Diamond Research at GE*, General Electric Company.
31 Suits, G.G. (1960) *The Synthesis of Diamond – A Case History in Modern Science*. Wiley.
32 Hall, T. (1970) Personal Experiences in High-pressure. *The Chemist*, **47**, 276–9.
33 Shigley, J.E., Fritsch, E., Koivula, J.I., Sobolev, N.V. *et al.* (1993) The Gemological Properties of Russian Gem-Quality Synthetic Yellow Diamonds. *Gems and Gemology*, **29** (4), 231.
34 Van Bocksteal and Van Royen (1996) *Antwerp Facets*, August, 55.
35 Fritsch, E., Conner, L. and Koivula, J.I. (1989) A Preliminary Gemological Study of Synthetic Diamond Thin Films. *Gems and Gemology*, **25** (2), 86.
36 Shigley, J.E., Fritsch, E., Stockton, C.M., Koivula, J.I. *et al.* (1986) The Gemological Properties of the Sumitomo Gem-quality Synthetic Yellow Diamonds. *Gems and Gemology*, **29** (4), 192.
37 O'Donoghue, M. (1997) *Synthetic, Imitation and Treated Gemstones*. Reed Educational and Professional Publishing.
38 Shigley, J.E., Fritsch, E., Reinitz, I. and Moon, M. (1992) An Update On Sumitomo Gem-Quality Synthetic Diamonds. *Gems and Gemology*, **28** (2), 116.
39 Chapman, L. (1980) *Diamonds In Australia: The Fields and The Prospectors*, Bay Books Ltd.
40 General Electric Company official website – http://www.ge.org
41 Welbourn, C.M., Cooper, M. and Spear, P.M. (1996) De Beers Natural Versus Synthetic Diamond Verification Instruments. *Gems and Gemology*, **32** (3), 156.
42 De Beers Industrial Diamond Product Chart (1988) *Indiaqua*, p. 160.
43 Koivula, J.I. and Fryer, C.W. (1984) Identifying Gem-Quality Synthetic Diamonds: An Update. *Gems and Gemology*, **21** (3), 147.
44 Nassau, K. (1990) Synthetics of the 80s. *Gems and Gemology*, **1**, 57–9.
45 Wei Li, Kangi, H. and Wakatsuki, M. (1996) *Journal of Crystal Growth*, **160**, 78–86.
46 Shigley, J.E., Moses, T.M., Reinitz, I., Elen, S., Shane, F. *et al.* (1997) Gemological Properties of Near-Colorless Synthetic Diamonds. *Gems and Gemology*, **33** (1), 51.

47 Press release issued by General Electric Company, 10 July 1990 and October 1991.
48 Bovenkerk, H.P., Bundy, E.P., Hall, H.T., Strong, H.M. et al. (1959) Man-made diamond. *Nature*, **184**, 1094–8.
49 Tolansky, S. and Sunagawa, I. (1959) Spirals and other growth forms of synthetic diamonds; a distinction between natural and synthetic diamonds. *Nature*, **184**, 1526–7.
50 Tolansky, S. and Sunagawa, I. (1959) Interferometric studies on synthetic diamonds. *Nature*, **185**, 203–4.
51 Anderson, B.W. (1990) *Gem Testing* (10th edition, rev. by E.A. Jobbins), Butterworth & Co Ltd.
52 Shigley, J.E., Fritsch, E., Reinitz, I. and Moses, T.M. (1995) A Chart for the Separation of Natural and Synthetic Diamonds. *Gems and Gemology*, **31** (4), 259.
53 Shigley, J.E., Fritsch, E. and Reinitz, I. (1993) Two Near-Colourless General Electric Type IIa Synthetic Diamond Crystals. *Gems and Gemology*, **29** (3), 186.
54 Shigley, J.E., Fritsch, E., Stockton, C.M., Koivula, J.I. et al. (1987) The Gemological Properties of the De Beers Gem-Quality Synthetic Diamonds. *Gems and Gemology*, **23** (4), 200.
55 Sosso, F. (1995) Some Observations on a Gem-Quality Synthetic Yellow Diamond Produced in the Region of Vladimir (Russia). *Journal of Gemmology*, **24** (5), 365.
56 Serway, R.A. (1996) Physics for Scientists and Engineers: with Modern Physics, 4th edition. Saunders.
57 Editorial (1994) Characteristics of natural and synthetic diamonds observed in cathode luminescence photographs. *Four Seasons of Jewellery*, **116**, 37.
58 Loudon, R. (1983) *The Quantum Theory of Light*, 2nd edition. Oxford University Press.
59 Born, M. and Wolf, E. (1987) *Principles of Optics: Electromagnetic Theory of Propagation, Interference, and Diffraction of Light*, 6th edition. Cambridge University Press.
60 Sobel, M.I. (1987) *Light*. University of Chicago Press.
61 Moses, T.M., Reinitz, I., Fritsch, E. and Shigley, J.E. (1993) Two Treated-Color Synthetic Red Diamonds Seen in the Trade. *Gems and Gemology*, **29** (3), 186.
62 Moore, C. (1971) *Atomic Energy Levels as Derived from the Analysis of Optical Spectra*. US National Bureau of Standards.
63 Sandström, A.E. (1957) Experimental Methods of X-ray Spectroscopy: Ordinary Wavelengths. *Handbuch der Physik*, **30**, 78–245.
64 Warren, B.E. (1990) *X-ray Diffraction*. Dover Publications.
65 Winick, H. and Doniach, S. (1980) *Synchrotron Radiation Research*. Plenum.
66 (1993) Beware: Russian Sells Man-Made Diamonds. *Mazal U'Bracha* **9** (53), 58–61.
67 Costan, J. (1993) Slow start for Chatham diamonds. *Diamonds International*, November/December, 71–4.
68 C3 Inc. Press release, 22 November 1996.
69 Bruton, E. (1978) *Diamonds*. 2nd edition, N.A.G. Press, London.
70 Rooney, M.T., Welbourn, C.M., Shigley, J.E., Fritsch, E. et al. (1992) De Beers Near Colourless-to-Blue Experimental Gem-Quality Synthetic Diamonds. *Gems and Gemology*, **29** (1), 41.

71 Sobolev, N.V., Efimova, E.S. and Pospelova, L.N. (1981) Native iron in diamonds of Yakutia and its paragenesis. *Soviet Geology and Geophysics*, **22** (12), 18–21.
72 Hodgkinson, A. (1996) Observation of Magnetism in Synthetic Diamond. *Gems and Gemology*, **32** (3), 154.
73 Kane, R.E. (1992) Graining in Diamond. *International Gemological Symposium Proceedings 1992*, Gemological Institute of America, pp. 219–35.
74 Hodgkinson, A. (1996) *Gems and Gemology*, **32** (3), 151.
75 Editorial (1994) CIBJO Moves on Disclosure of Gem Treatments, *Jewellery World*, **6**, 16.
76 Borgmans, P. (1996) *Antwerp Facets*, August issue.
77 Brozel, M.R., Evans, T. and Stephenson, R.F. (1978) Partial dissociation of nitrogen aggregates in diamond by high temperature – high-pressure treatments. *Proceeding of the Royal Society of London, A*, **361**, 109–27.
78 Liddicoat, R.T. (1996) Editorial: Opening Pandora's Black Box. *Gems and Gemology*, **32** (3), 153.
79 Hodgkinson, A. (1995) Magnetic Wand – Synthetic diamond detector. *Rapaport Diamond Report*, May 5, 34–5.
80 Argyle Diamonds Polished Sales Division (1997) Argyle Diamonds' Pink Diamond Tenders (1985–1996). *The Australian Gemmologist*, **19** (10), 415–18.
81 Collins, A.T. (1978) Migration of nitrogen in electron irradiated type Ib diamond. *Journal of Physics C: Solid State Physics*, **13**, 1417–22.
82 Collins, A.T. (1990) Optical centers in synthetic diamond – a review (R. Messier, J.T. Glass, J.E. Butler and R. Roy, eds, *Proceedings of the Second International Conference on New Diamond Science and Technology*, **23–237**, 667.
83 More information can be obtained on H3 centres in diamonds in numerous texts on diamonds and gemstones; however the author recommends *The Nature of Diamonds* (G. E. Harlow, ed.) (Cambridge University Press, 1998).
84 Collins, A.T. (1982) Colour centres in diamond. *Journal of Gemmology*, **18** (1), 37–75.
85 Collins, A.T. and Stanley, M. (1985) Absorption and luminescence studies of synthetic diamonds in which the nitrogen has been aggregated. *Journal of Physics D: Applied Physics*, **18**, 2537–45.
86 Collins, A.T. and Spear, P.M. (1982) Optically active nickel in synthetic diamonds. *Journal of Physics D: Applied Physics*, **15**, L183–87.
87 Von Bolton, W. (1911) *Elektrochem*, **17**, 971.
88 Derjaguin, B.V., Fedoseev, D.V., Lukyanovich, V.M., Spitzen, B.V. *et al.* (1968) Filamentary diamond crystals. *Journal of Crystal Growth*, **2**, 380–84.
89 Eversole, W. (1962) Patent Nos. 3030187 and 3030188. Synthesis of diamond. US Patent Office.
90 Angus, J.C. and Hayman, C.C. (1988) Low-pressure growth of diamond and diamond like phases. *Science*, **241**, 913–21.
91 Matsumoto, S., Sato, Y., Kamo, M. and Setaka, N. (1982a) Vapor deposition of diamond particles from methane. *Japanese Journal of Applied Physics*, **21** (4), L183–85.
92 Matsumoto, S., Sato, Y., Tsutsumi, M. and Setaka, N. (1982b) Growth of diamond particles from methane-hydrogen gas. *Journal of Materials Science*, **17**, 3106–12.
93 Derjaguin, B.V. (1976) *Dokl Akad Nauk SSSR*, **231**, 333–5.

94 May, P.M. (1995) CVD diamond – a new Technology for the Future? *Endeavour Magazine*, **19** (3), 101–6.
95 Simon, F.E. *et al.* (1961) *Low Temperature Physics: Four Lectures*. Pergamon.
96 Davies, P. (ed) (1989) The New Physics. *The Journal of Low Temperature Physics* (semi-annual), 268–88.
97 Rosenberg, H.M. (1963) *Low Temperature Solid State Physics: Some Selected Topics*. Clarendon Press.
98 Sittig, M. (1963) *Cryogenics: Research and Applications*. Van Nostrand.
99 White, G.K. (1987) *Experimental Techniques in Low-Temperature Physics*, 3rd edition. Clarendon Press.
100 Ashfold, M.N.R., May, P.W,. Rego, C.A. and Everitt, N.M. (1994) *Chemical Society Review*, **23** (21).
101 Spear, K.F. and Dismukes, J.P. (1994) *Synthetic Diamond: Emerging CVD Science and Technology*. Wiley.
102 Nicholson, E.D., Baker, T.W,. Redman, S.A., Kalaugher, E. *et al.* (1996) Young's Modulus of Diamond-Coated Fibres and Wires. *Diamond Related Materials*, **5**, 658–63.
103 Hay, R. (31-5-1995) Accelerated Commercialization of Diamond-Coated Round Tools and Wear Parts. Advanced Technology Project – project information.
104 Partridge, P.G., Ashfold, M.N.R., May, P.W. and Nicholson, E.D. (1995) The Effective Chemical Vapour Deposition Rate of Diamond. *Journal of Material Science*, **30**, 3973–82.
105 Robins, L., Plano, M.A., Moyer, M.D., Moreno, M.A. *et al.* (1994) The defect microstructure of thick homoepitaxial diamond films, *Advances in New Diamond Science and Technology* (S. Saito, N. Fujimori, O. Fukunaga, M. Kamo *et al.*, eds) p. 251.
106 Plano, M.A., Moyer, M.D., Moreno, M.A., Black, D. *et al.* (1994) Properties of thick homoepitaxial diamond films, in Diamond, SiC, and Nitride Wide-Bandgap Semiconductors. *Materials Research Society Proceedings* (C. Carter Jr, G. Gildenblat, S. Nakamura, R. Nemanich, eds), **339**, 307.
107 Black, D., Burdette, H., Plano, M.A., Moyer, M.D. *et al.* (1995) Observation of individual defects in a homoepitaxial diamond film. *Proceedings of the Fourth International Symposium on Diamond Materials*.
108 Krause, R. (18-3-1996) Chips Today, Diamond Skillets Tomorrow? *Investors Business Daily*, **4**.
109 Adiar, J. and Singh, R. (1995) *Science Daily Magazine*.
110 Graebner, J.E., Reis, M.E., Seibles, L., Hartnell, T.M. *et al.* (1994) *Physical Review, B,* **50** (6), 3702.
111 Streckert, G.G., Franch, H.G. and Collin, G. (1988) Schmuckstiene und Verfahren zu iher Herstellung. Bundesrepublik Deutschland, Deutsches Patentamt Offenlegungsschrift, Nummer DE 3708171A1.
112 May, P.W., Rego, C.A., Trevor, C.G., Williamson, E.C., Ashfold, M.N.R. *et al.* (1994) Deposition of Diamond Films on Sapphire: Studies of Interfacial Properties and Patterning Techniques. *Diamond and Related Materials*, **3**, 1375.
113 Koivula, J.L. (1987) Gem News – Cubic Zirconia coated by synthetic diamond? *Gems and Gemology*, **23** (1), 52.
114 Fritsch, E. (1997) Gem News – Synthetic diamond thin film jewelry. *Gems and Gemology*, **33** (2), 143–4.
115 Bachmann, P.K. (1995) Thermal properties of C/H-C/H/O-, C/H/N-, and C/H/X-

Grown Polycrystalline CVD Diamond. *Diamond and Related Materials*, **4**, 820–26.
116 Skokov, S., Weiner, B. and Frenklach, M. (1994) *Journal of Physical Chemistry*, **98**, 7073.
117 Skokov, S., Weiner, B. and Frenklach, M. (1995) *Journal of Physical Chemistry*, **99**, 5616.
118 MacInnes, D. (1973) *Synthetic Gem and Allied Crystal Manufacture*. Noyes Data Corporation.
119 Field, J.E. (1979) *The Properties of Diamond*. Academic Press.
120 (1966) Patent No. 3,238,019, to Stanford Research Institute, US Patent Office.
121 (1968) Patent No. 3,401,019, to El Du Pont de Nemours and Company, US Patent Office.
122 Garrett, D.R. (1970) Patent No. 3,499,732, US Patent Office.
123 (1972) Patent No. 3,632,242, to US Administrator to NASA, US Patent Office.
124 Garrett, D.R. (1972) Patent No. 3,659,972, US Patent Office.
125 McKenzie, D.R., Muller, D., Pailthorpe, B.A., Wang, Z.H. *et al.* (1991). Properties of tetrahedral amorphous carbon prepared by vacuum arc deposition. *Diamond and Related Materials*, **1**, 51–59.
126 (1977) Patent No. 1,473,313, to Harwell Atomic Energy Research Centre, UK Patent Office.
127 Patent No. 1,485,364, to Harwell Atomic Energy Research Centre, UK Patent Office.
128 John, P. (1996) *WG & MS Newsletter*, **10**.
129 For a description of the invention, design and use of the diamond anvil, suggested reading is Chapter 12 of Hazen, R.M. *The New Alchemists: Breaking Through the Barriers of High-pressure* (New York: Random House, Time Books, 1993), 1st edition.
130 Regueiro, M.N., Monceau, P. and Hodeau, J.L. (1992) Crushing C(60) to diamond at room temperature. *Nature*, **355**, 237–9.
131 Baum, R. (1992) High-pressure Changes C(60) to Diamond. *Chemical & Engineering News*, **5**.
132 Meilunas, R.J., Chang, R.P.H., Lui, S. and Kappas, M.M. (1991) Nucleation of diamond films on surface using carbon clusters. *Applied Physics Letters*, **59**, 3461–3.
133 Krätschmer, W., Lamb, L.D., Fostiropoulos, K. and Huffman, D.R. (1990) Solid C_{60}; a new form of carbon. *Nature*, **347**, 354–8.
134 Heiney, P.A., Fischer, J.E., McGhie, A.R., Romanow, W.J. *et al.* (1991) Orientational ordering transition in solid C_{60}. *Physical Review Letters*, **66**, 2911–14.
135 (1995) Patent No. 5,449,491, US Patent Office.
136 Visscher, G.T., Nesting, D.C., Badding, J.V. and Bianchoni, P.A. (1933) *Science*, **260**, 1496.
137 Brown, M.W. (4-6-1993) Tossing Plastics into an Oven, Scientists Cook up Diamond. *New York Times*, sec. A, p. 1.
138 Hemley, R.J., Mao, D., Shen, G., Badro, J. *et al.* (1997) *Science*, **276**, 1242–45.
139 Patent No. 5,449,491, US Patent Office.
140 Lang, A.R., Vincent, R., Burton, N.C. and Makepeace, A.P.W. (1995) Studies of Small Inclusions in Synthetic Diamonds by Optical Microscopy, Microradiography and Transmission Electron Microscope. *Journal of Applied Crystals*, **28**, 690.

141 Arima, M., Nakayama, K., Akaishi, M., Yamaoka, S. *et al.* (1993) Crystallization of diamond from silicate melt of kimberlite composition in high-pressure and high-temperature experiments. *Geology*, **21** (11), 968–70.
142 Browne, M.W. (1989) *New York Times*, 25 August, D18.
143 Bradley, D. (1993) What's harder than diamond? *New Scientist*, 22.
144 Bundy *et al.* (1996) *Carbon*, **34**, 141–53.
145 Suzuki, Lang (1976) *Journal of Crystal Growth*, **34**, 29–37.

Index

Adamas Gemological Laboratory, 91, 99–100
Adhesion – of diamond thin films, 131, 135, 136, 140
ADR, *see* Anomalous Birefringence
Alpha-alpha-alpha-trichlorotolulene, 147
Aluminium, 66, 139
Amethyst, 140
Anomalous Birefringence, 76, 103
    cause of, 103
    in natural diamonds, 103
    in synthetic diamonds, 103
Antoine Lavoisier, *see* Lavoisier, Antoine
Appophyllite, 140
ASEA, 31–5, 45, 128
    first synthesis, 33
    scandiamant DB, 50
    split sphere device, 32, 45
    synthetic diamonds, 33
Atmosphere, 4, 7, 11, 16, 21, 22, 23, 25, 32, 35, 51, 107, 119, 128, 130, 144, 149
    definition of, 21
    for diamond synthesis, 21, 22, 25, 27–9, 38–9
    measurement of, 11
Atomic Weight, 83, 85
    of Carbon, 15, 83
    of Lead, 83
Averani, G., 3

Baltzar Von Platen, *see* Von Platen, Baltzar
Bar, 21, 22, 142
Bachmann, 142
BARS apparatus, 45, 83, 107, 119, 123, 143
    application of, 45, 83, 119, 123
    description of, 45
Band gap, 62–3, 64, 65, 108
    in diamond, 63, 64, 65, 108
    theory of, 62–3, 65
BC8, 22
Belt apparatus, 10, 34, 35, 36, 39, 40, 42, 43, 52, 60, 143, 150
    description of, 39–41
    half, 39
    invention of, 35
    reaction cell, 32
Berman, R., 27
Berman-Simon Line, 27, 70
Beryl, 140
Bianconi, Patricia A., 147
Birefringence, 76, 103
Blu–tack, 114
Blue fluorescence, 80, 83, 84, 85, 100, 102, 104, 109
    to cathode rays, 82
    to long wave ultra-violet, 100, 102, 104, 109
    to short wave ultra-violet, 80
    to X-ray, 83–4
Blue light, 63, 102
Boron, 64–5, 112, 113
    as additive in synthetic diamonds, 65, 87, 98, 141, 142
    atomic number, 64
    in diamond, 64–5, 87
Boyle, Robert, 3
Bridgman, Percy, 7–13, 21, 22, 26, 33, 128
    at Harvard, 9–12
    discoveries, 9, 12, 26
    early work, 9
    gasket, 10, 60
    massive support, 12
    Nobel prize, 9, 12
    pressure measuring system, 11
    tapered anvils, 10–1, 33
Brilliant cut, 74, 88, 107, 116, 117
British Museum, 5
Bovenkerk, H. P., 33
Brodie, 3
Brown series, 102, 104, 109
Buckminsterfullerene, 3, 15, 16, 146
    discovery of, 3, 15, 16

in diamond synthesis, 146
in Fullerite, 146
structure of, 16, 17
Buckyballs, *see* Buckminsterfullerene
Bundy, Francis, 33, 35
Bragg, L., 18

Canon Diablo meteorite, 5, 25
Calcium oxide, 6
Cape Series, 77, 102, 103, 104, 105, 141
Carbon, 3–4, 5, 6, 7, 15–6, 17, 21, 22, 26, 27, 28, 32, 39, 52, 59–61, 66, 70, 99, 131–2, 145–6, 147
   allotropes of, 3, 15, 130, 146
   amorphous, 3, 115, 129
   atomic number, 15
   atomic weight, 15, 83
   bonding of, 4, 7, 21, 26, 59, 62, 75, 127, 132, 134
   chemistry of, 15, 26, 29, 60–1, 64, 66, 71, 127, 129, 132, 145–6
   isotopes, 16, 66, 99
   phases of, 5, 23, 24, 27, 28, 70, 115, 129, 134, 139
   sources of, 15, 127, 129
   study of, 15–16
   thermal properties of, 24, 127, 131
   triple point 21, 27
   valency, 15
Carbonado, 19, 43
Carbon dioxide, 3, 6, 21
Carbon nitride, 150
Carnegie Institute, Washington, 40, 149
   funding, 40
   high pressure research, 149
Cast iron, 6, 16, 53
Catalysts, 29–30, 39, 47, 52, 60–1, 66–7, 70, 72, 79, 88, 92, 117, 129, 150
   gaseous, 130–1, 146
   metals used as, 29, 47, 66, 70, 76, 93–4, 103, 117
   use of, 29, 39, 47, 66, 70, 71
Cathode rays, 81–2, 83
   discovery of, 82
   production of, 82, 83
Cathodoluminescence, 5, 73, 81–2
   colours of, 82
   importance of, 5, 82
   in identification, 81–2
   in natural diamonds, 82
   in synthetic diamonds, 82
   patterns in, 82
Celsius, 23
Celsius, Anders, 23
Chatham Created Gems Inc., 51, 68, 79, 118

Chemical vapour deposition, 9, 66, 127, 129–30, 132, 133, 134, 135, 136, 137, 138, 142, 145
   apparatus, 130–1
   invention of, 127
   process, 129–30, 132–3, 135, 136, 142, 143
   substrates, 131, 134, 135, 137, 142
   use of hydrogen, 129, 130, 131, 142
Cheney, J. E., 33
Chromium, 47
Citrine, 140
Clarity, 59, 91, 96, 109, 117
Clerici solution, 73
Cleavage, 5, 18–20, 52, 92
Clinopyroxene, 25
Coatings, 50, 53, 72, 127–8, 129, 133–4, 135, 136, 138, 140–2, 145
Cobalt, 29, 35, 70, 79, 94, 145
Coes, Loring, 21, 22, 25
Coesite, 22, 25
Collins, 5
Coloured diamonds:
   blue, 64–5, 66, 76, 79–80, 86, 93, 95, 96, 97–9, 103, 111, 113, 141
   brown, 76, 77, 83, 87, 98, 109, 111, 116, 119, 138, 140
   green, 76, 77, 83, 84, 87, 96, 98, 100, 119, 122
   pink, 78, 88, 116, 118
   red, 88, 98, 115–18
   yellow, 50, 51, 62, 63–4, 66, 69, 73–4, 76, 77, 79, 88, 93–100, 102, 103, 109,110, 111, 118–20, 141
Conduction – heat, 24, 139
Convection – heat, 24, 39
Cordite, 6
Corundum, 9, 19, 141
Copper, 24, 32, 55, 94, 102, 135, 143, 144–6
Covalent bonding, 15, 18, 21, 24, 26, 62, 145
Crookes, Sir William, 6, 82
Crystal, 5, 6, 18–9, 25–6, 32, 39, 45, 73, 76, 79, 82, 92–3, 98, 119, 128, 134, 141, 149–50
   graphite, 18–19, 73, 92, 93, 131–2, 141
   natural diamond, 5, 71–2, 73, 92
   synthetic diamond, 32, 35, 45, 50–4, 61–72, 76, 79, 82, 97, 107, 119, 128, 131–2, 141, 145–6
Crystallography, 18, 98, 136
Crystallume, 138
Cube, 32, 46, 67, 68, 70, 79
Cube-octahedron, 68–70
Cubic press, *see* Ultra-press
Cubic system, 18, 19, 20, 76
Cubic faces, 67, 68, 69, 70, 136

Cubic zirconia, 75, 91, 105, 140
Curl, Robert, 16
Cutting wheel, 74
CVD, see Chemical vapour deposition

De Beers, 39–40
  DiamondSure™, 105–8
  DiamondView™, 108–9
  early research, 39, 47
  high pressure apparatus, 39, 50
  legal battle with GE, 39, 40
  research laboratory, 39, 50
  synthetic diamonds, 68, 76–9, 82–4, 87–8, 91, 92, 93, 95–9, 105, 110–13
  synthetic grit products, 50, 54–5
DeCarli, P. S., 144
Density, 18, 19, 24, 54, 75, 146
Derjaguin, B. V., 128–9, 131
Desch, C. H., 7
Diamond, natural:
  age, 25
  birefringence, 76, 103–4
  cathodoluminescence, 82
  cleavage of, 5, 19–20, 52, 92
  coating of, 72, 141–2
  crystals of, 5, 71–2, 73, 92
  dispersion of, 20, 75
  diaphaneity, 3, 20
  electroluminescence, 84
  formation, 25
  fluorescence, 100, 104
  graining, 71, 94
  hardness, 7, 19, 20, 51, 74, 91
  inclusions, 25, 26, 89, 92, 94, 150
  metal in, 94
  morphology of, 17–21, 59, 71–2
  origin, 3, 16, 25–6, 48
  pink, 78, 88, 116, 118
  phosphorescence, 79
  refractive index, 20, 75
  specific gravity, 19, 20, 73
  thermoluminescence, 85
  twinning, 19, 71, 72
  types, 19, 20, 61–5
Diamond, synthetic:
  advantages over natural, 52
  applications for, 51–5, 74, 135–7
  birefringence, 76, 103–4
  blue, 64–5, 66, 76, 79–80, 86, 93, 95, 96, 97–9, 103, 111, 113, 141
  brown, 76, 77, 83, 87, 98, 109, 111, 116, 119, 138, 140
  cathodoluminescence of, 82
  cleavage of, 19, 20
  conditions for synthesis, 16, 22, 24, 27–30, 35, 45, 47, 50, 60–1, 143–50
  crystal faces of, 67–9
  dispersion of, 20, 75
  electroluminescence, 84
  faceting behaviour, 51, 74, 95, 96
  first synthesis, 32, 35, 128
  fluorescence of, 99–101, 104
  grading of, 96
  graining, 94–6
  grit, 37, 47–50, 65, 129
  green, 76, 77, 83, 84, 87, 96, 98, 100, 119, 122
  hardness, 20, 51, 74, 75, 127
  identification of, 48, 69, 72, 73–5, 82, 84, 91–6, 97–104, 105–14, 141–2
  inclusions in, 60, 67, 73, 74, 87, 88–9, 91, 92–4, 96, 104, 117, 150
  isotopically pure, 66, 79, 84, 86, 93, 95, 99
  mono-crystals, 50, 53, 54, 59–61, 65–9, 72, 74, 75, 79, 82, 129, 139, 141
  morphology, 52, 59–61, 67–71, 71–2
  near-colourless, 50, 61, 65–6, 68, 75, 78–80, 82, 84–6, 91, 93, 95–6, 97–103, 111, 112
  phosphorescence of, 79–80, 83, 84, 109
  production levels, 29, 44, 47–51, 53–4, 82, 129
  red, 88, 98, 115–18
  refractive index, 75, 95, 140
  specific gravity, 73–4
  spectrum of, 67, 68, 76, 78, 87–8, 101–3, 122
  thermoluminescence, 84–5
  twinning, 67, 136
  types, 61–5, 65–7
  yellow, 50, 51, 62, 63–4, 66, 69, 73–4, 76, 77, 79, 88, 93–100, 102, 103, 109, 110, 111, 118–20, 141
DiamondSure™, 105–8
  principal of, 105–6
  results from, 107
  use of, 106–8
DiamondView™, 108–9
  principal of, 108
  results from, 109
  use of, 108–9
Diamond anvil, 21, 22, 146
  limitations of, 22
Diamond coatings, 50, 53, 127–30, 133–4, 135–6, 138–9, 142
  applications of, 133–7
  on cubic zirconia, 140
  on natural diamond, 141–2
  on sapphire, 140
Diamond colour master stones, 101, 110, 111, 112

Diamond mining, 25, 150
Diamond probe, 84–5, 91, 97, 105, 139, 142
Diamond sensor, 113–4
Diamond wand, 113–4
Diamond thin films, 125–42
   cause of, 75
   compared to diamond films, 134, 139
   composition of, 134–140
   description of, 134, 136, 138–40
   dispersion, 20, 40, 75
   dissolution, 72
   DLC, 133–4, 135, 139, 140
   in diamond, 20
   in jewellery, 142
   measuring of, 75
   production of, 129–34, 142, 143–50
Dodecahedral faces, 67, 68, 69, 70, 72, 98
Dodecahedron, 19, 20, 67, 68, 71, 72, 98
Double refraction, *see* Birefringence
Du Pont, 49, 144

E-type, 26
Eclogite, 16, 25, 26
Electrical conduction, 5, 12, 24, 64, 139
   of diamonds, 24, 64
Electroluminescence, 84
   in natural diamonds, 84
   in synthetic diamonds, 84
Electromagnetic spectrum, 23, 74, 76, 78, 81,
   82, 85, 88, 143
Electrons, 15, 23–4, 61–5, 88, 94
Emerald, 51, 140, 141
Emerald cut, 74
Emplacement, 26
Energy:
   absorption, 63, 64, 65, 76, 85, 88
   barrier, 28–30, 132–3
   gap, 62–5, 108
   heat, 7, 21, 23, 24, 26, 28–9, 61, 84, 91,
      130–2, 135, 136, 139
   levels, 21, 23–4, 28–9, 63–5, 130, 132–3
Enhancement, 64, 68, 105, 115, 116
   colour, 116, 119–23, 141
Epigenetic inclusions, 92
Eversole, W. G., 128

Faceting, 51, 59, 74, 75, 95
Fahrenheit scale, 23
Ferromagnetism, 94
Fiveling, 72
Florence Academy, 3
Fluorescence:
   blue, 80, 83, 84, 85, 100, 102, 104, 109
   green, 79, 100, 104, 109, 116, 118, 120
   in natural diamonds, 100, 104

   in synthetic diamonds, 99–101, 104
   orange, 78, 79, 100, 116, 118
   patterns in, 99–101
   red, 84, 116
   to long wave ultra-violet, 78, 79, 99–100,
      104, 108, 116, 118, 120
   to short wave ultra-violet, 78, 79, 99, 100,
      104, 108, 116, 118
   yellow, 78, 79, 99–101
Fluoride, 141
Flux, *see* Catalysts
Fracture, 19, 20, 50, 92, 115, 141
Fraunhofer, Joseph, 74
Fraunhofer lines, 74–5
Freeman, J. H., 145
Frequency, 63, 65, 81, 85, 131, 143
Fullerite, 146–7

Gamma rays, 143
Garnet, 25, 94, 140
Gasket – non-extruding, 10, 60
Gemological Institute of America, 61, 66, 68,
   86, 92, 96, 110, 115, 118, 119, 123,
   138
   findings of, 61, 68, 69, 76, 83, 84, 85, 88,
      92, 93, 115–16, 118, 122
Gem Trade Laboratory, 66, 68, 76, 83, 96, 98,
   118, 123
   research by, 67, 68, 73, 74, 76, 78–9, 82, 83,
      95, 99, 115–19, 138–42
General Electric, 12, 21, 31, 37, 38–41, 42, 45,
   47, 50, 53–4, 66, 68, 73, 74, 128
   isotopically pure diamonds, 66, 79, 84, 86,
      93, 95, 99
   legal battle with De Beers, 39–40
   patents, 47
   Project Superpressure, 32–9, 42, 128
   synthetic diamonds, 67, 75–80, 83–4, 86, 91,
      92–3, 95–6, 99
   synthetic grit products, 53
Getters, 66, 67, 71
GIA, *see* Gemological Institute of America
Graphite, 3, 15–18, 19–21, 22, 24, 26
   discovery of, 3, 16
   optical properties of, 17, 18, 20
   physical properties of, 3, 17, 18, 19–21
   specific gravity of, 4
   structure of, 3, 17–8
Graining:
   in natural diamond, 71, 94
   in synthetic diamond, 94–6
Green fluorescence, 79, 100, 104, 109, 116,
   118, 120
   to cathode rays, 82
   to long wave ultra-violet, 116, 118, 120

Index   163

to short wave ultra-violet, 78, 79, 80, 100, 104, 109, 116, 118, 120
to X-rays, 83
Grenz rays, 83

Hall, H. Tracy:
  at General Electric, 40–2
  belt apparatus, 35
  Brigham Young University, 40
  megadiamond Corporation, 40–3, 137
  on making diamonds, 48
  tetrahedral Press, 40–2
Hand held spectroscope, 74, 76, 78, 101, 116, 117, 122, 123
Hannay James, Ballantyne, 4
Hardness, 3, 7, 19–20
  differences in, 74, 140
  directions of, 74
  of diamond, 19–20, 52–3, 74, 140
  scales, 19
Harvard University, 9, 12
Haske, Martin, 99–100
Heat conduction, 24
Heat convection, 24
Heat sinks, 24
Heat Treatment, see Treatment annealing
Herschel, W., 85
Hershey, J. W., 6
Hertz, G. L., 82
High pressure, 4, 9–13, 15, 21–2, 25, 26, 51, 79, 119, 128, 129, 130, 146
  for diamond synthesis, 21–2, 27–8, 32, 35, 45
  measurement of, 11–12, 21–2
  to bend diamond, 149
Hydrocarbons, 4, 16, 129
  in diamond synthesis, 129–32, 133, 136
Hydrogen, 4, 23, 129, 131, 140, 141, 142, 149
  in diamond synthesis, 129, 131, 142

Ice, 9, 23
Inclusions:
  in synthetic diamonds, 60, 67, 73, 74, 87, 88–9, 91, 92–4, 96, 104, 117, 150
  in natural diamonds, 25, 26, 89, 92, 94, 150
  in identification, 73, 91–4, 96, 104
  magnetic, 93, 94, 96, 104
Industrial diamonds, 48–50, 53–5
  uses, 48–9, 51–3
  natural diamond grit, 52
  synthetic diamond grit, 50, 51–5
Indentation hardness, 19
Infrared, 20, 64, 68, 119, 143
  spectroscopy, 68, 85–88, 92, 98, 116, 117, 123

radiation, 20, 85, 143
  discovery of, 85
  in identification, 86–8, 98, 11677, 117
Ion implantation, 145
Isometric, 19, 20
Isotopes, 16, 66, 99
Iron, 3–7, 16, 29, 32, 45, 50, 53, 66, 87, 89, 94, 112–113, 117

Jameson, J. C., 144
James Ballantyne Hannay, see Hannay, James Ballantyne
Jewellery, 51, 65, 91, 96, 99, 105, 107, 112, 114, 115, 118, 123, 140
  diamond thin films in, 142
Jessup Ralph. S., 26

Kennedy, George, 21, 42
  De Beers witness, 40
Kelvin, 1st Baron, 23
Kelvin scale, 23
Kimberlite, 25, 72, 150
Knoop scale, 19
Kroto Harry, 16
Kyanite, 140

Lamproite, 25, 72
Latent heat, 21, 23
Lattice, 16, 18–19, 22, 26, 27, 61–5, 66, 75, 76, 83, 85, 95, 102, 117, 118, 128, 131–4, 145, 149
Lavoisier, Antoine, 3
Lazare Kaplan, 74
Lead, 6, 16, 83
Liquid carbon, see Carbon, phases of
Lithium, 4
Loring Coes, see Coes, Loring
Low pressure, 127–31, 136, 143, 145
Luminoscope, 81, 119
Lundblad, Erik G., 33
Lustre, 19–20

Macle, 19, 72
Magnesium, 4
Magnetism, 93, 94, 96, 104, 112–14
  cause of, 93, 94, 96
  in identification, 102–14
  in natural diamond, 93, 94, 96
  in synthetic diamond, 93, 94, 96, 112–14
  of Blu-tack, 114
Manganese, 29, 45, 47
Manhattan Project, 9
Massive support, 12, 35, 40
Masters, see Diamond colour master stones
Megabar, 22

164 The diamond formula

Megadiamond Corporation, 40–3, 137
Mercury, 12, 21, 78, 82, 128
Metastability of diamond, 22
Methane, 16, 129, 142
   in diamond synthesis, 129, 142
Michel, H., 82
Microwaves, 85, 130, 131, 142, 143
Microwave generator, 130, 131
Microscope examination, 5, 94–7, 103, 104, 116
Miller indices, 67–70, 98, 136, 150
Mohs' scale, 7, 19, 20
Moissan, Frédéric-Henri, 5
Moissanite, 6, 7, 91
   naming of, 7
   synthetic moissanite, 6, 7, 91
Molybdenum sulphide, 3
Morphology, 52, 59, 75, 150
   description of, 67–9
Manipulation of, 69–71
   in identification, 71–2, 75
   in natural diamonds, 71–2, 150

N-type conductors, 74
Nanometre, 18, 78, 85, 109
Naphthalene, 30
National Bureau of Standards, 12, 21
National Research Development Corporation, 145
National Institute of Standards & Technology, 143
Natural diamonds, *see* Diamonds, natural
Near-colourless synthetic diamonds, 50, 61, 65, 66, 68, 75, 76, 78–80, 82, 84–6, 91, 93, 95–103, 111, 112
Nerad, H. A., 38, 39
Newton, Isaac, 74
Nickel, 29, 45, 55, 66, 70, 76, 78, 79, 87, 89, 92, 94, 103, 104, 117–20, 123
   solubility of carbon in, 29–30
Nitrogen, 5, 61–5, 76, 78, 79, 87, 96, 98
   in natural diamond, 5, 61, 62, 63
   in synthetic diamond, 65–6, 70, 74, 78, 79, 87, 96, 98, 107
   atomic number, 61
Nitrogen getters, 66
   metals used as, 66
Non-extruding gasket, 10
Norton Company, 12, 25, 33, 135
Nucleation, 52, 61, 71, 137
Nyf, 72
Neodymium, 112

O'Donoghue, M., 115
O'Ruff, 6

Octahedral faces, 19, 20, 68, 69, 79, 150
Octahedral habit, 20, 52, 67, 68, 70, 95
   in natural diamonds, 19, 20, 71, 72
   in synthetic diamonds, 52, 67, 68, 70, 71, 95
Octahedron, 19, 45, 67–8, 71
Opal, 140
Oppenheimer, J. Robert, 9
Olivine, 25, 92
Optical properties:
   diamond thin films, 134, 136, 138–40
   graphite vs. diamond, 20
   in identification, 20, 75, 95, 103–4, 140
   synthetic diamond, 20, 75, 76, 95, 99–101, 103–4, 140
   see also specific materials
Orange fluorescence, 78, 79, 100, 116, 118
   to long wave ultra-violet, 116, 118
   to short wave ultra-violet, 78, 79, 100, 116, 118
   to thermoluminescence, 85
Oxygen, 3, 5, 6, 12, 21, 23

P-type, 26
Palladium, 29, 30
Parsons, Charles Algernon, 6–7
Paraffin, 4, 16
Pascal, 21
Patents, 31, 39, 43, 47, 50, 128, 136, 140, 142, 143, 144, 146, 147
Peanuts, 16
Pearl, 59, 71, 82
Pennsylvania State University, 129, 147
Peridotite, 25–6
Pink diamonds, 78, 88, 116, 118
   natural, 88
   treatment of, 118
   spectrum of, 78
Pipestone, 10
Phase diagram, 27, 28, 70
Phase transition, 9, 22, 23, 24, 59, 60
Phases of carbon, 5, 15, 24, 28, 129, 139, 143
Phosphorescence:
   in synthetic diamonds, 79–80, 83, 84, 109
   in natural diamonds, 79
Physical properties, *see specific materials*
Plasma, diamond synthesis, 130–33, 142–3, 144, 145
   In diamond synthesis, 130–3, 142–3, 144, 145
Platinum, 29, 30, 47
Polariscope, 103, 104, 138
Polaroid, 103
Polycrystalline diamond, 53, 55, 133, 136, 137, 138, 142, 146
Polymers, 16, 136, 147

in diamond synthesis, 147
Potassium, 4
Potassium nitrate, 3
Pressure, 4, 21–2, 27–8, 32, 35, 42
   for diamond synthesis, 22, 27, 29
   high, 11, 21–2, 25, 26, 51, 79, 119, 128, 129, 130, 146
   low, 127–31, 136, 143, 145
   measurement of, 11–12, 21–2
   multiplication of, 10
   units, 21
Project superpressure, 33–9, 128
Propane, 16, 129
Protogenetic inclusions, 92
Proceedings of Royal Society, 5
Pyrophyllite, 10, 33, 39, 42, 60
Pyroxene, 25

Quantum theory theory, *see* Band gap theory
Quartz, 6, 7, 22, 130, 131
Quenching, of diamond, 146–7
QUINTUS, 33

Radiant cut, 88, 116, 117
Radiation, 20, 23, 24, 63, 73, 74, 77–80, 81–4, 85–8, 108, 110, 120, 143, 149
Rankine scale, 23
Red diamonds, 88, 98, 115–18
   features of, 115–18
   treatment of, 115–8
Red fluorescence, 84, 116
   to long wave ultra-violet, 116
   to short wave ultra-violet, 116
Refractive index, 9, 20, 75, 95
   description of, 20, 95
   for spectrum, 75
   of diamond, 20, 75, 96
   of diamond films, 140
   of DLC, 140
Refractometer, 140
Riedl, G., 82
Rossini, Fredrick D., 26

SAS2000 Spectrophotometer, 101, 102, 109–12
Sapphire, 5, 140
SC4, 22
Scale:
   Celsius, 23
   Farenheit, 23
   Kelvin, 23
   Knoop, 19
   Mohs, 7, 19
   Rankine, 23
Scheele, K. W., 3

Scratch test, 5, 7
'See though' technique, 75, 91
Seeds, 35, 39, 50, 59–60, 61, 77, 128, 145
Semi-conductors, 66, 135, 139, 141
Shock-wave process, 144–5
Silicon, 25, 129, 131, 138–9, 147
Silicon carbide, *see* Moissanite
Silicon nitride, 136
Simon, R., 27
Slater, John C., 9
Smalley, Rick, 16
Soapstone, 32
Sodium, 4, 75, 147
Sodium light, 20, 75
Solar spectrum, 75
Solvents, *see* Catalysts
Specific gravity, 4, 19–20, 73–4
Spectrum, 101–4
   brown series, 102, 104, 105, 109
   cape series, 77, 102, 103, 104, 141
   nickel related, 76, 78, 103, 104, 117, 118, 120, 123
   nitrogen related, 78, 104, 117
Solar, 75
Spectrophotometer, 62, 76, 77, 101, 102, 109–13, 116, 119, 120, 121
Spectroscope – hand-held, 74, 76, 78, 101, 116, 117, 122, 123
Spinel, 6, 7
   early confusion with diamond, 7
Story-Maskelyne, Nevil, 5
Strong, H. M., 33, 35, 39
Strain, *see* Anomalous birefringence
Strontium titanate, 140
Subduction, 16
Substrates, 130–4, 136–40, 142, 143–4, 145–6
Sugar, 5, 16
Sumitomo, 49, 50, 61, 68, 91, 141
   policy, 50
Synthetic diamonds, 61, 68, 69, 73, 74, 76, 78–9, 83–4, 86, 93, 95–8, 100, 112, 118
Sumicrystal™, 50, *see also* Sumitomo, synthetic diamonds
Symmetry, 18, 98
Syngenetic inclusions, 92
Synthetic Diamond Sensor, 113–14

Tabby extinction, *see* Anomalous birefringence
Targiono, C. A., 3
Tairus, 51
Tantalum, 32
Tapered anvils, 10–11, 33, 35, 40
Temperature:
   gradient of, 5, 50, 60

for synthesis, 24, 27, 29, 30, 32, 35,45, 128, 130, 144, 145, 147
scales, 23
Tennant, Smithson, 3
Tetrahedral:
   bonding polymer, 147
   lattice, 18, 26, 75, 128, 131, 132, 134, 145–6, 147
Tetrahedral press, 40, 41, 42, 143
   invention of, 40–2
Tetrahedron, 42, 147
Thermal-conductivity meter, 84, 139
Thermite, 32, 135
Thermoluminescence, 84–5
   in natural diamonds, 85
   in synthetic diamonds, 84–5
Thermo-probe, 84, 139
Thomson, William, 150
Titanium, 66
Tourmaline, 140
Tracy Hall, see Hall, H. Tracy
Transmittance, 101, 102, 110, 111, 112, 113
Transparency, 3, 19, 29, 83, 95, 122, 138, 142, 143, 144, 147
Trapezohedral faces, 68, 69, 70, 72
Treatments, 115–23
   annealing, 107, 116, 118–19
   colour treatment, 115–18, 118–23
   irradiation, 83, 115, 116
   Russian techniques, 118–23
Tungsten, 131, 135, 143
Tungsten carbide, 16, 35, 45, 53, 55, 107, 135
Twinning, 19, 20, 67, 71, 136
   in natural diamonds, 19, 71, 72
   in synthetic diamond, 67, 136
Type I diamonds, 19, 20
Type Ia, 61, 64, 83, 86, 87, 88, 102, 123, 142
Type Ib, 61, 64, 65, 76, 79, 84, 86, 87, 88, 96, 97, 101, 102–3, 107, 118, 120, 123
Type II diamonds, 19, 20, 64, 107
Type IIa, 65, 68, 76, 79, 84, 86, 88, 96, 97
Type IIb, 64, 65, 80, 86, 87, 96, 97, 139

Ultra High-Pressure Limited, 50
Ultra-press – six anvil, 42, 43
Ultra-violet light:
   long wave, 78, 79, 99–100, 104, 108, 116, 118, 120
   short wave, 78, 79, 99, 100, 104, 108, 116, 118, 120

in diamond grading, 99
in identification, 104
synthetic diamond reaction, 99–101, 104
natural diamond reaction, 100, 104
Union Carbide, 128
Unit cell, 18

Vapour deposition, see Chemical vapour deposition
Verdict – GE/De Beers case, 40
Von Bolton, W., 128
Von Platen, Baltzar, 21, 31–7
   at ASEA, 31–3
   inventions of, 31

Water, 6, 19, 23, 33, 73, 85, 114, 140, 143
   high pressure forms, 9, 12, 22
Wavelength, 20, 62, 64, 75, 77, 78, 81, 83, 85, 100, 108–10, 120–2, 138
Wentorf, Robert Jnr., 33, 35

X-rays, 74, 82–4, 88, 143
   discovery of, 82–3
   production of, 82, 83
   luminescence to, 83–4, 88–9
   verification using, 5, 17, 18
X-ray diffraction patterns, 5, 17, 18, 37

Yellow cathodoluminescence, 82
Yellow diamonds, 50–1, 62–4, 66, 69, 73–4, 76, 77, 79–88, 93, 95, 96, 102–3, 109–11, 118–20, 121, 141
   causes of colour, 63–4
   types, 62, 64, 74, 79, 86, 87, 95, 102, 103, 118
   in nature, 62–4, 110, 141
   to X-ray, 83–4
Yellow fluorescence, 78, 79, 99–101
   to long wave ultra-violet, 78, 79, 104
   to short wave ultra-violet, 78–9, 99–101, 104, 120
Yellow phosphorescence, 84
Yellow zoning, 97–9, 116–17

Zenolyth, 125
Zenocryst, 25
Zirconia, see Cubic zirconia
Zirconium, 66, 140
Zoning, 71, 96, 118, 119–20
   colour, 87, 95, 97–9, 116, 117, 118, 119–20
   fluorescent, 99–101, 116